Waterfowl Ecology

TERTIARY LEVEL BIOLOGY

A series covering selected areas of biology at advanced undergraduate level. While designed specifically for course options at this level within Universities and Polytechnics, the series will be of great value to specialists and research workers in other fields who require knowledge of the essentials of a subject.

Recent titles in the series:

Saltmarsh Ecology	Long and Mason
Tropical Rain Forest Ecology	Mabberley
Avian Ecology	Perrins and Birkhead
The Lichen-Forming Fungi	Hawksworth and Hill
Social Behaviour in Mammals	Poole
Physiological Strategies in Avian Biology	Philips, Butler and Sharp
An Introduction to Coastal Ecology	Boaden and Seed
Microbial Energetics	Dawes
Molecule, Nerve and Embryo	Ribchester
Nitrogen Fixation in Plants	Dixon and Wheeler
Genetics of Microbes (2nd edn.)	Bainbridge
Seabird Ecology	Furness and Monaghan
The Biochemistry of Energy Utilization in Plants	Dennis
The Behavioural Ecology of Ants	Sudd and Franks
Anaerobic Bacteria	Holland, Knapp and Shoesmith
An Introduction to Marine Science (2nd edn.)	Meadows and Campbell
Seed Dormancy and Germination	Bradbeer
Plant Growth Regulators	Roberts and Hooley
Plant Molecular Biology (2nd edn.)	Grierson and Covery
Polar Ecology	Stonehouse
The Estuarine Ecosystem (2nd edn.)	McLusky
Soil Biology	Wood
Photosynthesis	Gregory

TERTIARY LEVEL BIOLOGY

Waterfowl Ecology

MYRFYN OWEN BSc, PhD
and
JEFFREY M. BLACK, BA, PhD
The Wildfowl and Wetlands Trust,
Slimbridge, Gloucester.

Blackie
Glasgow and London

Published in the USA by
Chapman and Hall
New York

Blackie and Son Limited,
Bishopbriggs, Glasgow G64 2NZ
and
7 Leicester Place, London WC2H 7BP

Published in the USA by
Chapman and Hall
a division of Routledge, Chapman and Hall, Inc.
29 West 35th Street, New York, NY 10001—2291

First published 1990

British Library Cataloguing in Publication Data

Owen, Myrfyn
Waterfowl ecology. — (Tertiary level biology).
1. Wildfowl
I. Title II. Black, Jeffrey M. III. Series
598.41

ISBN 0-216-92675-0
ISBN 0-216-92676-9 pbk

Library of Congress Cataloging-in-Publication Data

Owen, Myrfyn.
Waterfowl ecology/M. Owen and J.M. Black.
p. cm. — (Tertiary level biology)
Bibliography: p.
Includes index.
ISBN 0-412-02191-9. — ISBN 0-412-02201-X (pbk.)
1. Anatidae—Ecology. I. Black, J.M. II. Title. III. Series.
QL696.A52088 1990
598.4′1—dc20 89-15903
CIP

Typesetting by Thomson Press (India) Limited, New Delhi
Printed in Great Britain by Thomson Litho Ltd, Scotland.

Preface

Waterfowl are an attractive and well-studied group of animals. They are specialists at exploiting a particular habitat—wetlands. The wide diversity of wetland types, however, and the geographical range over which they are found, means that waterfowl are found throughout the world, from north of 80° north in the Arctic tundras and seas, to tropical rain forests and marine habitats. One of the most interesting features of the group is the extreme mobility of most of the species, which capitalise on seasonal variations in the abundance of food and especially changes in its availability to water-based birds.

Because of Man's interest in the group, there have been many books published on waterfowl. Most, however, adopt a species-by-species approach or deal with identification or aspects of Man's relationship with waterfowl.

Waterfowl Ecology illustrates evolutionary and ecological principles, using examples from this unique group. We hope to give the reader an understanding of basic ecological processes and relationships, as well as of the interactions between waterfowl and their environment. The examples include some classical works, but we have mainly concentrated on the most recent studies. We have consulted a very wide range and large number of references; the extensive bibliography included should enable the student to follow up subjects of interest.

Waterfowl are much affected by the activities of Man, and we have illustrated in several sections the ways in which these interactions have modified the birds' behaviour or ecology. In the last chapter, in particular, examples of the effects of Man's activities, both destructive and constructive, illustrate the critical nature of the interaction of Man with individual species and with the environment in general.

Our understanding of waterfowl has benefited from discussions with a large number of workers worldwide. However, we would like to mention in particular the late Sir Peter Scott who, through his establishment of the

Wildfowl and Wetlands Trust and his other activities throughout the world, has done more than anyone else to promote the understanding and appreciation of waterfowl. Half a century ago, he recognised that, from the most superficial to the deepest academic level, waterfowl could be used as an example to further environmental education and conservation. We hope that this book makes some contribution to that end.

M.O.
J.M.B.

The Wildfowl and Wetlands Trust

Founded in 1947 by Peter Scott, The Wildfowl and Wetlands Trust is a registered charity whose objectives are conservation, research, recreation and education. It operates eight centres around Great Britain where nearly a million visitors a year appreciate the sight of wild and tame waterfowl at close quarters. The Trust has a research team of 20 scientists who are responsible for monitoring wildfowl populations in the United Kingdom. The team carries out a large number of other biological projects specialising in population dynamics and long term studies.

Acknowledgments

Photographs courtesy of Wildfowl and Wetlands Trust (Figure 1.2), M. Owen/Wildfowl and Wetlands Trust (Figures 3.3 and 7.2), J.B. Blossom/Wildfowl and Wetlands Trust (Figures 5.1 and 7.5) and M.J. Brown/Wildfowl and Wetlands Trust (Figure 7.6).

Contents

CHAPTER ONE
INTRODUCTION

Wetland habitats are among the most productive of ecosystems and harbour a unique variety of plants and animals. This book is about the ecology of a group of wetland specialists; although the group is well defined it does contain species of a wide variety of morphological characteristics and habits.

In our treatment we have adopted an evolutionary approach, dealing with the functions and adaptive significance of the various traits. We have used as examples some of the most recent studies, chiefly those which are manipulative or experimental, testing hypotheses and predictions. Although the treatment is restricted to waterfowl, the group does serve to illustrate general principles of evolution and natural selection. For a more general treatment see Perrins and Birkhead (1983) and Welty and Baptista (1988).

1.1 What are waterfowl?

The term 'waterfowl' includes different birds in Europe and North America; for the purposes of this book we have adopted the North American usage, which includes in the definition only members of the family ANATIDAE—ducks, geese and swans. This definition is identical to the term 'Wildfowl' in the British usage; in Europe the term waterfowl includes a very large and diverse group of birds which are ecologically dependent on water at some stage of their life cycle.

The waterfowl, which are placed in the Order ANSERIFORMES, represent a very distinct group, with few links with other families. The most closely related birds, and the only other group which are placed in the same Order are the screamers ANHIMIDAE, a family of only three species of long-legged, hook-billed birds restricted in distribution to South America. The Flamingos are also rather closely related but are placed in a separate Order PHOENICOPTERIFORMES.

The Anatidae contains some 143 species and about 250 forms or

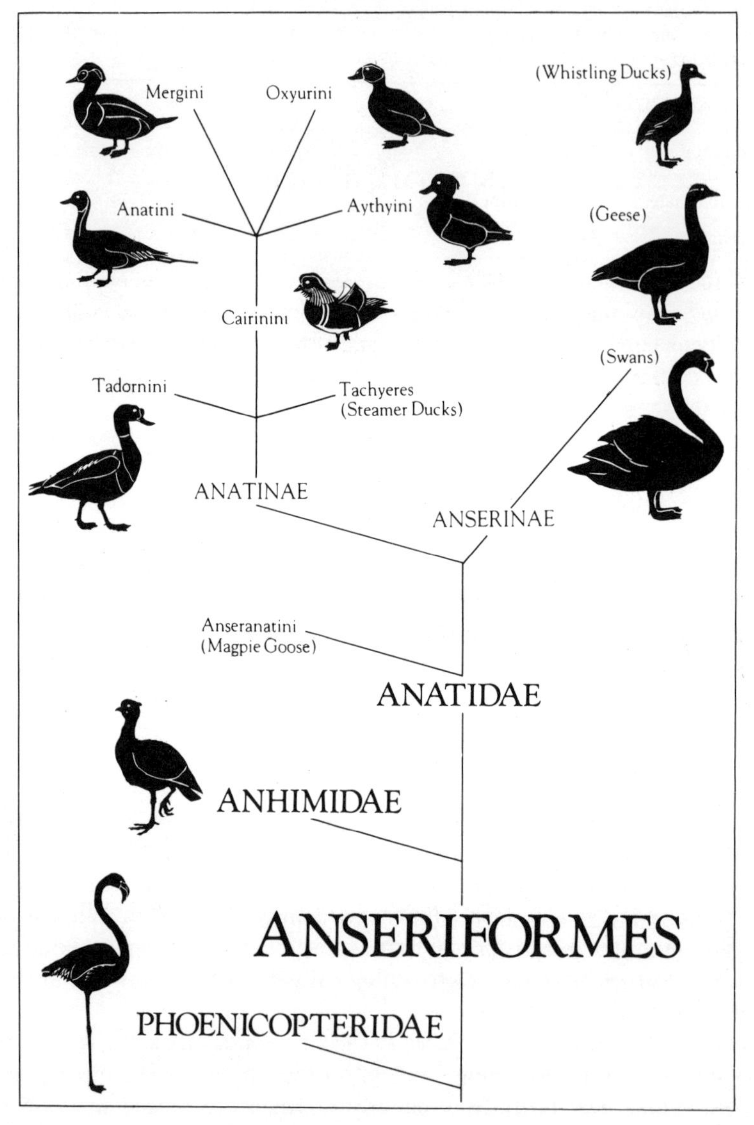

Figure 1.1 A simplified evolutionary tree of the waterfowl and their close relatives. This follows Johnsgard (1965) and Kear (1970). Drawn by J.B. Blossom and reproduced from Owen (1977a).

Table 1.1 Summary of the groups of the family Anatidae according to the classification of Johnsgard (1965), with slight amendments based on more recent work.

Family ANATIDAE Subfamily	Tribe	No. of genera	No. of species	Common name
Anseranatinae	Anseranatini	1	1	Magpie Goose
Anserinae	Dengrocygnini	1	8	Whistling Ducks
	Anserini	5	21	Swans and Geese
	Strictonettini	1	1	Freckled Duck
Anatinae	Tadornini	5	14	Sheldgeese/Ducks
	Trachyerini	1	3	Steamer Ducks
	Cairnini	9	13	Perching Ducks
	Anatini	5	40	Dabbling Ducks
	Aythyini	2	15	Pochards
	Mergini	7	18	Sea Ducks/Sawbills
	Oxyurini	4	9	Stifftailed Ducks
3	11	41	143	

subspecies. They have evolved from land birds, so the more terrestrial species show behaviour and life history traits which are considered primitive and the more aquatic ones more advanced (Kear 1970). Although there are debates about the relationships of individual aberrant species or genera, the family is split into eleven tribes, whose characteristics are described briefly below (for more detailed treatment see Johnsgard (1965), whose authority we have followed here). A summary of the relationships of the Anatidae and their close relatives is given in Figure 1.1.

The most primitive subfamily, the ANSERANATINAE, consists of a single tribe including only a single species—the Magpie Goose, a long-legged upright vegetarian species restricted to Australia.

The subfamily ANSERINAE consist of largely terrestrial long-legged birds divided into three tribes. The DENDROCYGNINI—the Whistling or Tree Ducks are tropical, with the eight species ranging from the Americas to Australia.

The ANSERINI include the geese and the swans. There are six species of swans, which are birds of subarctic and temperate regions of both hemispheres. The true geese are restricted to the northern hemisphere, are terrestrial for most of their life cycles and are almost entirely vegetarian grazers. An aberrant species—the Freckled Duck—is restricted to Australia and shows characteristics of geese and ducks, without the characteristic patterned plumage of the TADORNINI and is placed in its own tribe STRICTONETTINI.

The TADORNINI include the shelducks and sheldgeese, the latter being very similar in form and ecology to the true geese but are sexually dimorphic and share some behavioural features with the ducks. The shelducks are duck-like in apperance and represent an intermediate between them and the geese. A single genus consisting of three species of steamer ducks (two flightless) is placed in a separate tribe—the TRACHYERINI. All three species of steamer ducks occur in marine waters of South America and the Falkland Islands.

The perching ducks CAIRNINI are the tribe showing most variation in body size, from the relatively massive (up to 10 kg) Spur-winged Goose to the diminutive Pygmy Geese which, at about 300 g are among the smallest of the Anatidae. Most species are somewhat arboreal, nesting in tree holes or raised platforms. The group includes the ducks with the most spectacular plumage—the Mandarin Duck of Asia and the North American Carolina or Wood Duck.

The largest tribe is the ANATINI, the group known as dabbling ducks, which has 40 species, largely dimorphic ducks of shallow waters which range in distribution from the Arctic to the tropics. They walk well on land and gather much of their food whilst standing or swimming in very shallow water. Members of the tribe are found in all continents and in a wide range of fresh and brackish habitats.

The fifteen species in the AYTHYINI are diving ducks of predominantly fresh or brackish waters. Species occur in temperate and subarctic areas of the Northern Hemisphere and in Australasia. They are accomplished divers and collect most of their food underwater, to a depth of up to 6 m.

The MERGINI include the Eider Ducks which are sometimes given a tribe of their own; they are a group of four closely related sea ducks of exclusively Arctic and north temperate seas, where they forage by diving and have a diet made up largely of molluscs and crustaceans. The tribe also includes the scoters, goldeneyes and fish-eating sawbilled ducks, which inhabit relatively open, predominantly brackish or marine waters. They are mobile and expert divers and specialise on active invertebrate prey and, in the case of the sawbills, fish.

The OXYURINI or stiff-tailed ducks represent the group most adapted to an aquatic existence. They are round-bodied and have the legs set very far back on the body. This enables most efficient propulsion in the water but stifftails are very ungainly on land, where they seldom venture.

In the following chapters of this book we will attempt to find examples from as many of the groups and parts of the geographical range as possible. However, since most of the research has been concentrated on migratory species of the Arctic and north temperate regions, most of the examples come from the wealth of information on these groups.

1.2 Waterfowl and Man

Waterfowl have a long association with Man; aboriginal peoples, especially in Arctic regions, relied on them substantially for food. Most waterfowl are conspicuous, occur in open, treeless areas and often in large colonies, making them easy to exploit by egg-collecting. The majority of species moult their primary flight feathers simultaneously and become flightless for some weeks. This makes them vulnerable to being rounded up in large groups, a traditional activity in arctic North America, Iceland and Eurasia.

Waterfowl still form an important component of the diet of the native peoples of the Arctic, though now they are more commonly harvested by shooting, both in autumn and spring. This may have an impact on populations (see Chapter 6).

In temperate and southerly areas, waterfowl have traditionally been caught in a variety of traps and nets. The most sophisticated of these is the

Figure 1.2 An aerial view of the duck decoy at the Wildfowl and Wetlands Trust's Centre at Slimbridge. The decoy consists of a secluded pond set in woodland, from which lead four catching 'pipes'—tapering ditches over which are netted hoops. Note that the 'skate's egg' design allows at least one pipe to be used in any wind condition (the wind must blow down the pipe so that the ducks take off towards the catching area at the end).

decoy, still in use in north-west Europe to capture ducks for ringing. A decoy consists of 1–8 'pipes' constructed around a small secluded pond, usually in a wood. Each pipe consists of a tapering curved tunnel made up of netting stretched over hoops and bordered on the side away from the pond by a series of screens (Figure 1.2). Ducks gather to roost on the quiet pond and are lured into the pipes by a small, fox-like dog. Ducks on water will follow a fox in a 'mobbing response' and the decoyman uses his trained dog to capitalise on this behaviour. Decoys are most common in the Netherlands, where they were developed, and in Britain. An efficient decoy had a full-time operator catching up to several thousand ducks annually for the market.

Waterfowl are still hunted and eaten in temperate and subtropical wintering areas, though the primary motivation for present-day hunting is for sport rather than for food. In Eurasia the sport of waterfowl hunting (traditionally 'wildfowling', largely practised on the coast in Britain) is controlled by the restriction of quarry species and by the recognition of a close (non-hunting) season.

In North America, hunting regulations are extremely flexible and annually variable on the basis of the abundance of quarry or on a geographical 'sharing' of the harvest. Hunting seasons can vary from a few hours to several months, and species are regularly removed from and replaced onto the quarry list. The number of birds shot is controlled using 'bag limits', which can vary on a daily, individual or seasonal basis. In some areas hunters are encouraged, by a 'points system' to shoot some species (or sexes where there is a skewed sex ratio) according to their abundance or scarcity. The hunting kill is an important mortality factor for many species and this is considered in more detail in Chapter 6.

Several waterfowl species have been domesticated (Delacour 1954); the Egyptian Goose and the Greylag have been tamed in Egypt for at least 4000 years. Initially, no doubt, the eggs are collected from the wild and the young fattened in captivity, but selective breeding of captive stock has, in some species, resulted in a variety of forms which have been bred for different purposes.

It is not surprising that the most sedentary and adaptable species have made the most successful domestic ones. Among the geese, the Greylag has resulted in a number of forms, many pure white, for fattening, egg laying or even to act as 'watchdogs' or for fighting. The Swan Goose was domesticated by the Chinese centuries ago, though the present day domestic Chinese Goose is little different in plumage from its ancestor. The most extreme form is the very heavy African Goose which, nevertheless, has natural plumage characteristics.

The domestic Mallard is now found in a wide variety of forms, the most familiar of which is the white Aylesbury type which is the main breed used for the table. Less heavy strains developed as egg layers or as dual purpose breeds are more similar in plumage to the wild Mallard; the most extreme in form is the Indian Runner, which has a long neck and a near-upright posture.

The Muscovy Duck originates from South America and in its original plumage is black with a greenish or purplish sheen and with conspicuous black wattles around the base of the bill. It has been domesticated by the indigenous South Americans for centuries and present-day forms are mostly white or pied and the wattles are bright red.

No other species has been truly domesticated, but a number of species have traditionally been kept in captivity; the North American Indians used to collect eggs of Canada Geese and fatten them in pens. A stranger custom occurred in England, where the Mute Swan was semi-domesticated. Flocks were managed by the church or by companies and were marked with the owners' marks on the bill. In Abbotsbury, in Dorset, England, a colony of Mute Swans has been fostered and managed by monks for at least 600 years. Breeding swans were encouraged to nest and were protected, but the young cygnets were penned and fattened for the table. The 'swannery' is still run on traditional lines by a private estate, though the cygnets are now ringed and released rather than killed.

1.3 Waterfowl and wetlands

Waterfowl are among a number of groups of birds which, at least at some part of their life cycle are ecologically dependent on wetlands. The species which perhaps conforms least closely to this generalisation is the Hawaiian Goose which is, nowadays at least, found almost exclusively on the high altitude lava deserts of the Hawaiian Islands. It is one of the very few Anatidae that readily copulates on land but it betrays its ancestry since its pre-copulatory displays include ritualised bathing movements.

A wetland is defined in the *International Convention for the Conservation of Wetlands Especially as Waterfowl Habitat*, signed in Ramsar, Iran in 1971 (known as the *Ramsar Convention*) as follows:

> ...wetlands are areas of marsh, fen, peatland or water, whether natural or artificial, permanent or temporary, with water that is static or flowing, fresh, brackish or salt, including areas of marine water the depth of which at low tide does not exceed six metres.

The definition is clearly focused on the ecological requirements of waterfowl (6 m represents the normal limit of the deepest divers). Recent additions to the convention recognise that the number of waterfowl a wetland supports, which is the main criterion for the designation of sites under the convention, does not entirely describe the ecological or conservation value of that wetland, but waterfowl numbers are relatively easily assessed and provide a convenient indicator of the health of a wetland.

Wetlands are found in all regions of the world and there have been many, often complex, classifications. Although in this slim volume we cannot provide a comprehensive account we can give the reader an introduction to the complex interactions between waterfowl and their environment and, hopefully, encourage further reading from the bibliography. The importance of wetlands will become most evident in the sections on movements and migrations (Chapter 5) and conservation and management (Chapter 7). The most recent classification system for wetlands has been developed for the Ramsar Convention Sectretariat by Scott (1989) and represents the simplest workable classification. Here, we describe briefly the main types, merely to put the rest of the book in context and to make the reader aware of the diversity of the habitats used by waterfowl.

The simplest primary division is into fresh and salty waters. Many species of sea and steamer ducks inhabit marine habitats throughout the year and many others winter in saline waters and breed inland. An exclusively marine existence requires physiological adaptations, particularly the development of salt glands, which enable the excretion of excess salt; some of the sea ducks can drink only sea water over extended periods.

Shallow coastal wetlands, bays, straits and fjords, support some of the largest concentrations of waterfowl. Sea ducks remain at sea throughout the winter, diving for food over mollusc beds, especially the Blue Mussel *Mytilus edulis*. Estuaries, where the silt-laden waters of rivers meet the sea are very productive habitats, and probably represent the most important single type of wetland for waterfowl. Many species specialise on the intertidal area, feeding either on the limited range of vegetation there (Brent Goose) or on the plentiful invertebrates (many species of dabbling ducks and shelducks). Estuaries are among the most threatened of wetlands (Chapter 7).

Rivers provide habitats for waterfowl at all times of year, and several species, such as the torrent ducks of South America and the Blue Duck of New Zealand, have specialised in fast flowing upland waters. In the lowlands, rivers are prone to overflow their banks at times of high rainfall and the resulting floodlands are of prime importance to waterfowl. In Europe and North America many of the great rivers have been canalised and straightened and no longer flood. Only small patches of floodlands, mainly areas specially protected for waterfowl, remain. In the tropics, however, inland deltas and ephemeral waters are the most important habitats.

Shallow eutrophic lakes are often surrounded by extensive areas of marshland which support a wide range of breeding as well as wintering species. Some of these in inland South America and Africa are saline and remain open throughout the year. In the mid-latitudes of North America and Eurasia there are large areas of marshes and lakes which are very productive breeding grounds for ducks. The area known as the 'prairie pothole' region of central North America is often referred to as the 'duck factory', producing millions of ducks which migrate to the southern states in winter.

In the subarctic, the northern coniferous forests or **taiga** regions contain large areas of lakes and marshes, frozen in winter but providing breeding grounds for a wide variety of ducks, geese and swans. Among the most characteristic birds of these forests are goldeneyes, which nest in tree holes and rear their young on pools and marshes.

At higher latitudes, between the tree-line and the polar deserts in the Northern Hemisphere are vast areas of **tundra,** where migratory species nest in great numbers in the short Arctic summer. The tundra is not only highly productive of protein-rich vegetation in the short growing season, but the freshwater pools are excellent breeding grounds for insects, on which the young of most of the ducks flourish.

Artificial wetlands also provide valuable habitats in many regions. Reservoirs, created for drinking water, hydroelectric or irrigation schemes, supply still water where none existed before. Mineral workings often flood after extraction and provide important habitats of increasing value as they age. Sewage treatment ponds and settling pools for dredged wastes are also fertile feeding grounds.

Waterfowl quickly learn to capitalise on rich sources of food, and many of these have been created by Man for food production. These include rice and other cereal fields close to waterfowl roosts, exploited by large flocks of ducks. Geese and the smaller swans can travel many kilometres from roosting waters to arable fields if the food source is rich enough. Carnivorous species can be troublesome on fish or crustacean farms and some of the sea ducks prey upon molluscs in artificially seeded beds.

In short, wherever there is water, there are waterfowl to exploit the safety it provides and the food it produces. Many of the more mobile species depend on a wide variety of wetlands over a range of several thousand kilometres and international co-operation in research, legislation and practical management of wetlands is essential for the conservation of such populations (Chapter 7).

CHAPTER TWO

FOOD AND FEEDING ECOLOGY

This chapter deals with the relationship between waterfowl and their habitat, chiefly in winter when, for most species, there is the greatest differentiation between diets and habits. In the remainder of the book it will become evident that the search and competition for food is not only responsible for the movements and breeding behaviour of waterfowl, but also regulates populations and influences individual reproductive success. In this chapter we consider how closely related species coexist, their foraging routines, choice of food and the annual energy cycle. To indicate the diversity of adaptations among waterfowl, we adopt a comparative approach to illustrate the various accounts by examples from a wide variety of species and lifestyles.

2.1 Adaptive radiation

Species which coexist in the same environment, must have developed adaptations of structure or feeding habits which enable them to avoid competition in times of shortage. Species are able to coexist and exploit similar foods in times of plenty or at stages of the life cycle when there is no habitat limitation. Over time, waterfowl species have evolved to exploit the wide variety of niches available in the aquatic environment. We will argue that in times before Man's wholesale interference with the natural environment, it was in winter that competition was likely to be most severe, leading to adaptive radiation in feeding behaviour and morphology.

2.1.1 *Eurasian geese*

A good example of adaptive radiation of feeding methods and apparatus is found in the geese of North America and Eurasia. The five species which have traditionally wintered in Britain are adapted to feed on the different kinds of vegetation on the areas of open land that existed before Man felled the forests and introduced domestic grazing animals (Owen 1976). The

Figure 2.1 A diagrammatic representation of the place of five species of geese in the British landscape before deforestation. BA Barnacle Goose, BR Brent Goose, PF Pink-footed Goose, GL Greylag, GWF Greenland White-fronted Goose. The European Whitefront (EWF) may not have wintered in Britain in former times. Based on Owen (1976) and reproduced from Owen (1977a).

position of each species in the pre-deforestation landscape, as deduced from their morphological features and present habits, is shown diagrammatically in Figure 2.1.

The largest species is the Greylag, which used the *Scirpus* marshes in estuarine areas and inland fens where it fed on roots by digging. The Pinkfoot and White-front are intermediate in size; the former is a mobile bird of dynamic estuaries where it traditionally followed the pioneer grass zone. The White-front's present distribution matches the areas in western Britain and Ireland where there are bogs; it specialised in probing for roots and bulbils in the wet flushes among the peat. The Barnacle Goose has a short bill, which enables it, with a rapid pecking action, to graze very short swards of the exposed islands off the west coast. The Brent is a truly maritime species, feeding on Eelgrass *Zostera* in the intertidal zone and on the saline plants on saltmarshes. The five goose species vary in their mobility and gregariousness, and the variations match well their traditional habits although now they are on very similar, agricultural, habitats.

The conformation of the heads and bills of geese—their primary feeding apparatus—are well adapted to their winter foods; in the summer all species are grazers on tundra vegetation. Apart from the Greylag, they are in very similar breeding habitats, though in different geographical areas. This suggests that it was in winter that competition for food gave rise to the adaptive radiation in feeding apparatus and flocking behaviour. The evidence from what we know of the extent of traditional habitats in summer and winter in both Eurasia and North America, suggests that this is the case. The available breeding habitat in the vast Arctic tundra and taiga regions was sufficient for larger populations than could be supported by the limited open land in the largely forested temperate wintering range (Owen 1980a, pp. 130–134).

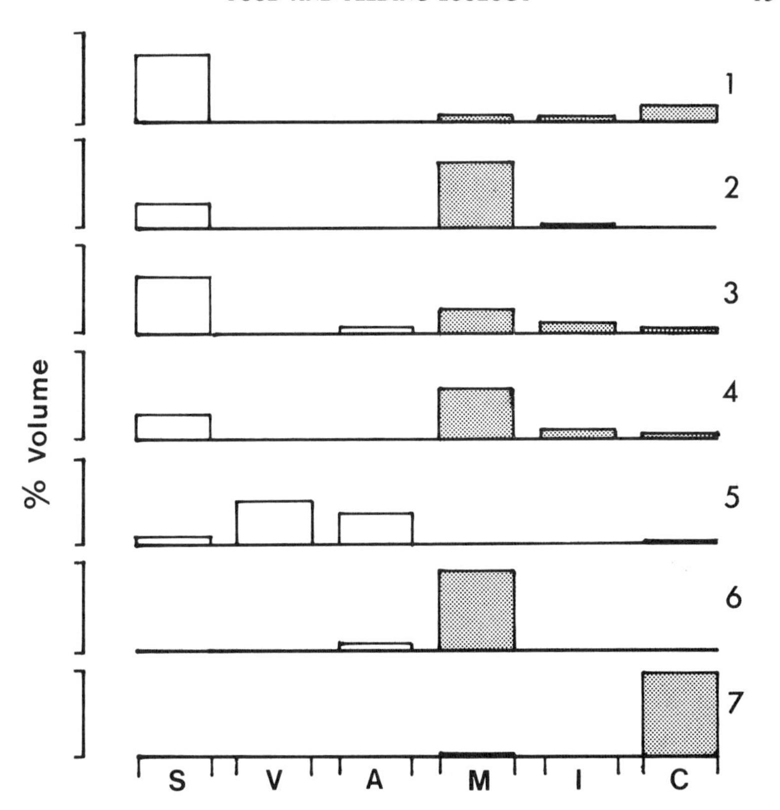

Figure 2.2 The percentage by volume of the main items in the diets of seven species of ducks on an estuarine habitat in south-east England, as determined by gut analysis. Foods: S seeds, V vegetative plant parts, A algae, M molluscs, I insects and C crustaceans. Animal items shaded. Species: 1 Mallard, 2 Pintail, 3 Teal, 4 Shoveler, 5 Wigeon, 6 Shelduck, 7 Goldeneye. Redrawn from Olney (1965).

2.1.2 *Sympatric ducks*

Throughout their life cycle, and in all parts of the world, a number of species of duck occur in the same waters and many of them have rather similar diets and feeding methods. The northern dabbling ducks provide a good example. Olney (1965) described the diets and suggested niche specialisation in dabbling ducks on British estuaries (Figure 2.2). There is a large amount of overlap in diet, but differences in feeding methods and morphological adaptations fit each species to a different niche.

The morphological and behavioural features of the same species were

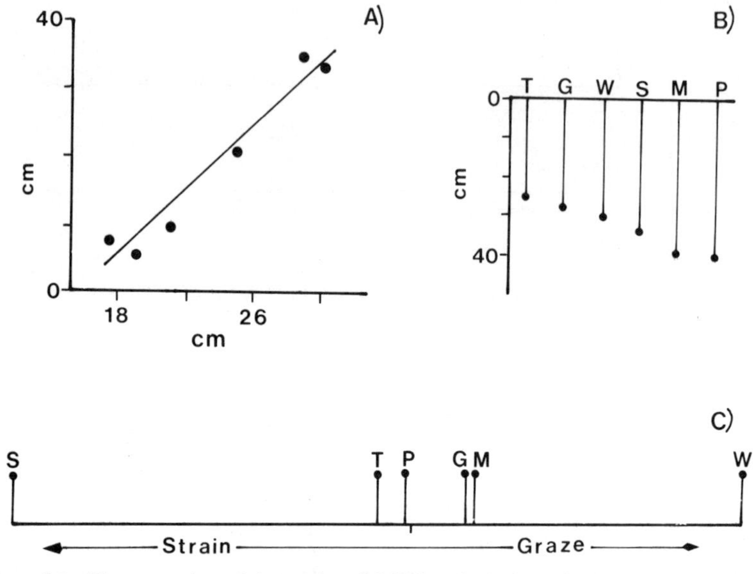

Figure 2.3 The separation of six species of dabbling ducks in a Finnish breeding area by morphology and feeding behaviour: a) correlation between the mean feeding depth in the habitat and neck length; b) maximum depth reached when up-ending; and c) differentation on eight measurements of bill morphology, indicating adaptations for staining and grazing. G—Gadwall, M—Mallard, P—Pintail, S—Shoveler, T—Teal and W—Wigeon. From Poysa (1983).

examined in more detail, using a multivariate approach, by Poysa (1983). The species lie on a gradient in terms of the length of the neck (maximum feeding depth) and bill characteristics which effectively enable them to exploit different parts of the habitat when there may be competition (Figure 2.3). Although all species can feed in a variety of ways, they adopt the range of feeding behaviour most suited to their morphological adaptations. Within the same depth range, differences in bill morphology, particularly the number of straining lamellae and the distance between them, allow some separation in terms of diet. The Shoveler is the extreme straining specialist, sieving out the smallest particles from the surface film. It has a similar neck length to the Wigeon, but that species has a short bill adapted for grazing and grubbing; the two species are almost completely separated in the diet they can exploit by virtue of the differences in their bill morphology.

Bill structure was examined in 22 species of North American waterfowl by Kehoe and Thomas (1986). The species, which ranged from the Snow Goose to the American Merganser, separated predictably according to the

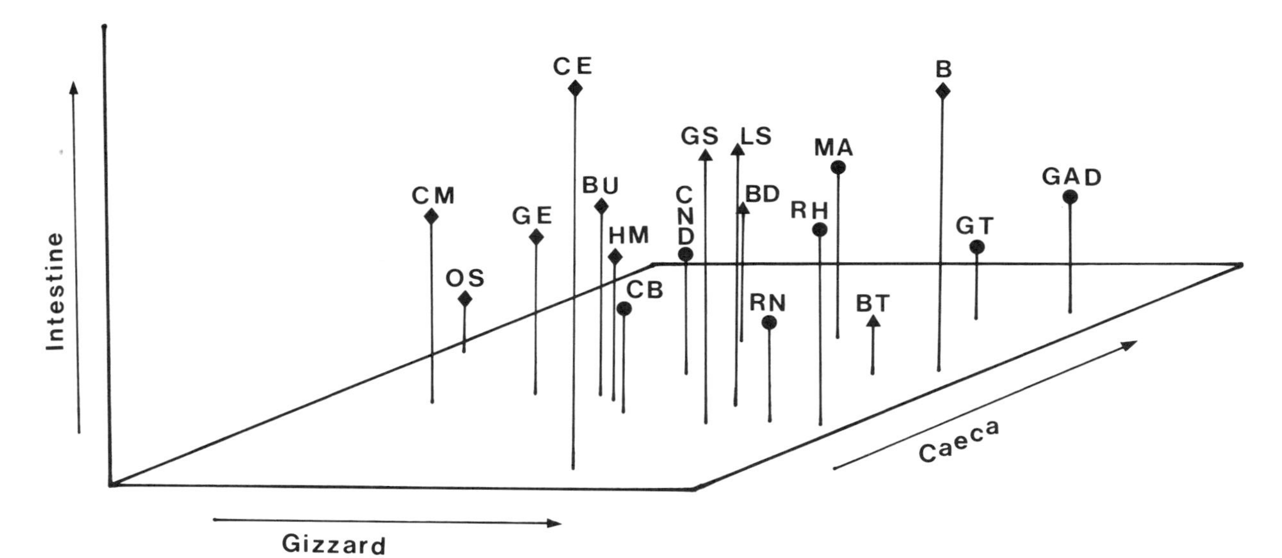

Figure 2.4 The separation of 18 species of North American Anatidae (discriminant function analysis) on the basis of digestive organ weights. Species codes: B—Brant, BD—Black Duck, BU—Bufflehead, BT—Blue-winged Teal, CB—Canvasback, CND—Canada Goose, CE—Common Eider, CM—American Merganser, GAD—Gadwall, GE—Goldeneye, GS—Greater Scaup, GT—Green-winged Teal, HM—Hooded Merganser, LS—Lesser Scaup, MA—Mallard, OS—Oldsquaw (Long-tailed Duck), RH—Redhead, RN—Ring-necked duck. Diet symbols: diamond—carnivore, triangle—omnivore, and dot—herbivore. From Barnes and Thomas (1986).

feeding habitats and foods they exploit. They were similarly separated in their internal feeding apparatus (gizzard, intestine and caeca—Barnes and Thomas (1986)—Figure 2.4). The Snow Goose has a heavy gizzard to cope with its fibrous diet, as do the other vegetarians Brant and Gadwall. The Oldsquaw (Long-tailed Duck), a carnivorous diver, has the smallest gizzard and a short intestine. As in their outward appearance, the Mallard and the Black Duck are similar in bill and gut morphology. The two species occupy a similar ecological niche in different parts of the continent and there are worries nowadays that as the Mallard spreads westwards it will increasingly come into competition with the Black Duck, to the detriment of the latter.

A number of species of sea ducks occupy shallow coastal waters. Although many of the species have similar diets, exploiting molluscs and crustaceans gathered from the sea bottom, they are effectively isolated by differences in their diving abilities. The Scaup reaches depths of 5 m, whilst the closely related Tufted Duck dives to around 7 m. The scoters can obtain food at up to 10 m, whilst the best diver, the Long-tailed Duck, reaches 20 m (Nilsson 1970, 1972). Among North American diving ducks, the size of the heart relative to body size, increases with diving depth; diving ducks as a group have relatively larger hearts than dabbling ducks and geese, and the Long-tailed Duck has much the largest heart of the group. This reflects the heavy oxygen demand of diving (Bethke and Thomas 1988).

In the breeding areas bird densities tend to be less than in winter, and the species use rather more similar habitats and relatively abundant food. In a study of breeding ducks in Finland an analysis of community ecology did not support the prediction that species show divergent feeding methods and habitats, and led to the conclusion that the community was not in a state of competitive equilibrium (Poysa 1984).

This is consistent with our argument that it is in winter that the crucial competition occurs in ducks as well as geese. The effect of this competition on the two groups is, however, quite different; the geese are separated in different habitats, whilst the ducks are able to coexist because they use different parts of the same environment. To avoid competition, the species have evolved differences in feeding apparatus, body morphology and physiological capability.

2.2 Exploitation of the habitat

The individuals which exploit their habitats and food supplies in the most profitable way benefit in terms of fitness, which affects breeding success or survival. Many studies have been carried out to determine how closely individuals' performance matches that expected by 'optimal foraging' theory (Krebs and Davies 1987). Hypotheses based on optimal foraging predict that an animal operates in relation to the costs and benefits of feeding in a particular way so that it maximises its gain from foraging. The most relevant measure of benefit is usually regarded as the net energy gained per unit time—the measure of 'profitability' used in the following account.

2.2.1 *Swans*

Swans traditionally fed by grazing on underwater vegetation, from a swimming position or by up-ending. The Mute Swan has a reach of more than a metre when up-ending, whereas the Bewick's and Whooper Swans that are sympateric with it in winter have to feed in shallower water. On water, swans are safe from land predators and feed at night as well as by day. At the Ouse Washes, England, Mute and Bewick's Swans begin feeding at sunrise and by midday about 80% of the flock are foraging. This level of activity continues into the night, to at least two hours after sunset (Owen and Cadbury 1975). Whooper Swans in Scotland spend only 16% of the 24 hours feeding in a situation where they have access to supplementary food, compared with 45% in a more natural situation (Black and Rees 1984).

2.2.2 *Geese and sheldgeese*

Geese are generally diurnal terrestrial feeders, spending the night in the safety of a roost on water or on mudflats or islands. They fly to nearby feeding grounds near first light and stay for most of the day. If the food supply is energy-rich and easily gathered they feed for a rather small proportion of the day and may return to the roost in midday. Some grazing species such as the White-fronted Goose may have to feed for up to 95% of the available daylight hours in midwinter (Owen 1972a). The feeding rhythm of some coastal species, such as Brent and Snow Geese, which can feed in safety at night, is related to tidal rather than light cycles (Charman 1980). The birds are, however, forced to adapt their behaviour to the feeding situation. Barnacle Geese staging in Norway adopt different patterns of foraging on undisturbed traditional islands as opposed to agricultural fields (Figure 2.5), because the latter are disturbed during the

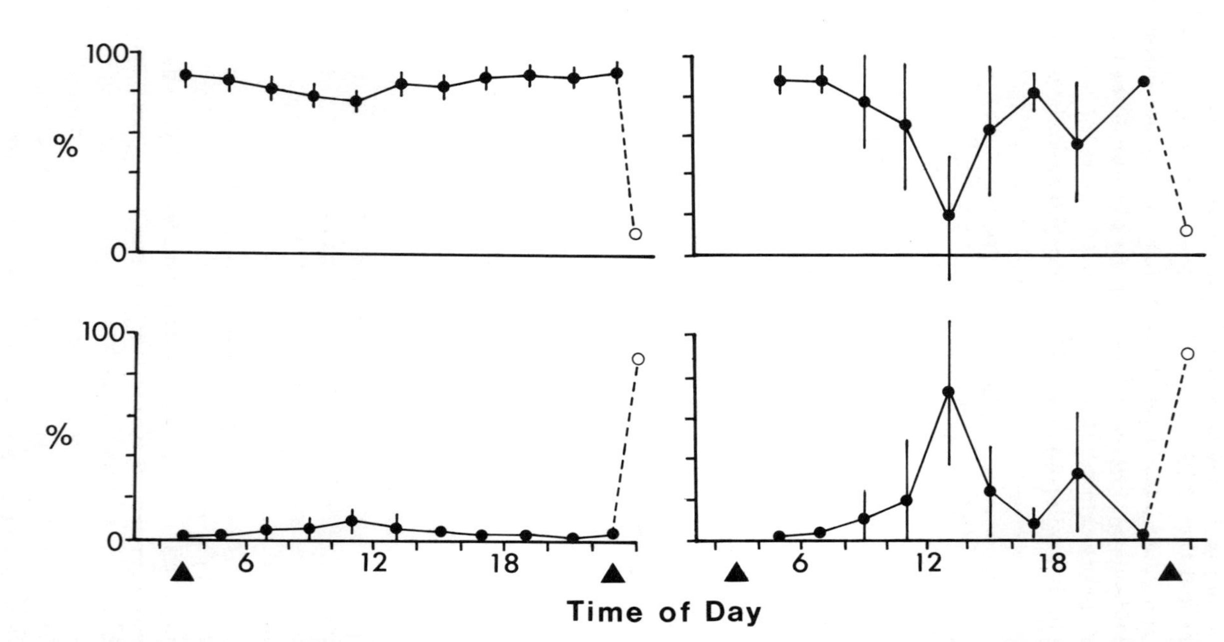

Figure 2.5 The diurnal activity patterns of Barnacle Geese at a Norwegian staging area on traditional outer island habitats (left pair of figures) and agricultural fields (right). The proportion of time spent feeding is the upper of each pair and the proportion resting the lower. From Black *et al.* (1990).

day and provide more nutritious forage. Disturbance was also the main factor determining the choice of feeding fields by White-fronted Geese (Owen 1972b). Similarly, Prins and Ydenberg (1985) failed to explain the choice of feeding site by Barnacle Geese in the Netherlands entirely by energetic considerations, but when they took the location of the feeding area into account and particularly the level of human disturbance, they did find that the birds were operating in the most profitable way.

The ecological similarity between true geese and the sheldgeese of South America is illustrated by detailed studies carried out on the Falkland Islands (Summers and Grieve 1982). Upland and Ruddy-headed Geese showed a diurnal feeding rhythm, and fed for 90% of the day. They selected good quality vegetation and had a very similar intake rate and digestive efficiency (25% of the energy content of food was retained) to true geese.

On a smaller scale, some elegant studies of food exploitation by geese have demonstrated how sensitive the birds are to changes in their environment. Brent Geese visiting a saltmarsh in the Netherlands in spring concentrate their feeding on the Saltmarsh Grass *Puccinellia maritima* and Sea Plantain *Plantago maritima* (Prins *et al.* 1980). The geese graze in large flocks and visit the same area of saltmarsh every 3–5 days (Figure 2.6).

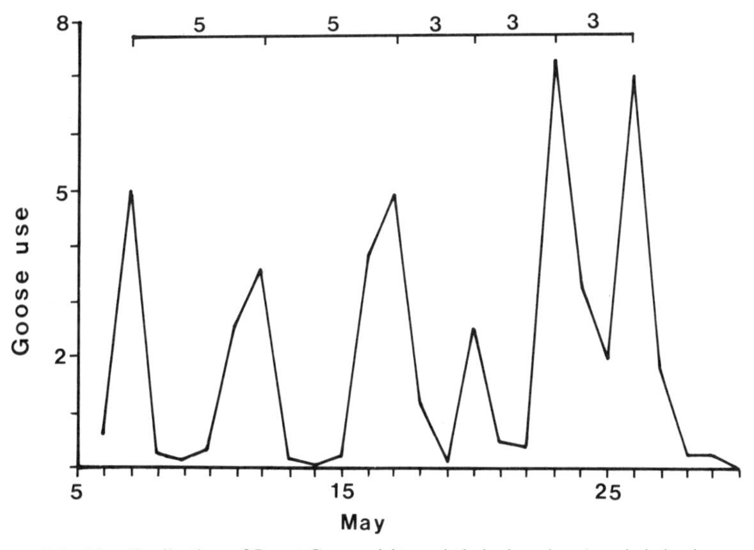

Figure 2.6 The distribution of Brent Goose visits and their duration (y axis is in thousands of goose minutes) on an area of saltmarsh in the Netherlands. Indicated along the top is the interval in days between the peaks of visitation. From Prins *et al.* (1980).

Clearly the peaks in goose visitation are very regular. Prins *et al.* also found that the length of the visit was related to the length of time since the last visitation had occurred, implying the geese were responding to the regrowth of the plants.

This study together with a neat clipping experiment added to the suggestive evidence that the geese were 'farming' the vegetation, returning to the same area only when sufficient time had elapsed to make another harvest a profitable proposition. Individual leaves and rosettes of *Plantago* were marked and some rosettes were subjected to different levels of clipping at different intervals to simulate defoliation by geese. The act of clipping stimulated the growth of the plant (up to a certain threshold level), so that the total amount of material produced by lightly grazed plants was almost double that in ungrazed vegetation. Clipping also concentrated growth in the upper (younger) leaves, whereas in ungrazed plants growth of all leaves proceeded at much the same rate. This had the effect of concentrating nutrients in the upper leaves, which also were more digestible to geese.

The most interesting finding of the clipping experiment was that the greatest productivity benefit was obtained when the plants were clipped every four days and when a third of the material was removed at a visit. This was exactly what the geese did, on average. This indicates how geese can manage both the quantity and quality of their food. The selective advantage of such behaviour is strong at a time when the birds are laying down fat reserves in preparation for migration and breeding. The amount of accumulated reserves has a direct effect on the likelihood of successful breeding (see page 31).

In the breeding areas, geese are present throughout the growing season of plants and have a unique opportunity to affect their own food supplies for that and future seasons. Studies at a breeding colony of Snow Geese at Hudson Bay, Canada showed that there was indeed a beneficial effect of grazing on overall productivity of saltmarsh vegetation. Grazing by geese speeds up the nitrogen cycle in the marsh; goose droppings provide nitrogen in a form which the plants can re-absorb and convert into new protein (Cargill and Jeffries 1984).

These studies also indicate possible feeding advantages that geese may get from being gregarious, both by feeding in flocks and nesting in colonies (see also page 84).

Other studies on the nesting grounds indicate how individual birds can benefit from their ability to monitor the habitat and harvest the food efficiently. In Spitsbergen, Barnacle Geese nest on offshore islands and during incubation breaks females fly to the nearby mainland to feed on the tundra plants as they emerge from the melting snow (Figure 2.7). Each day

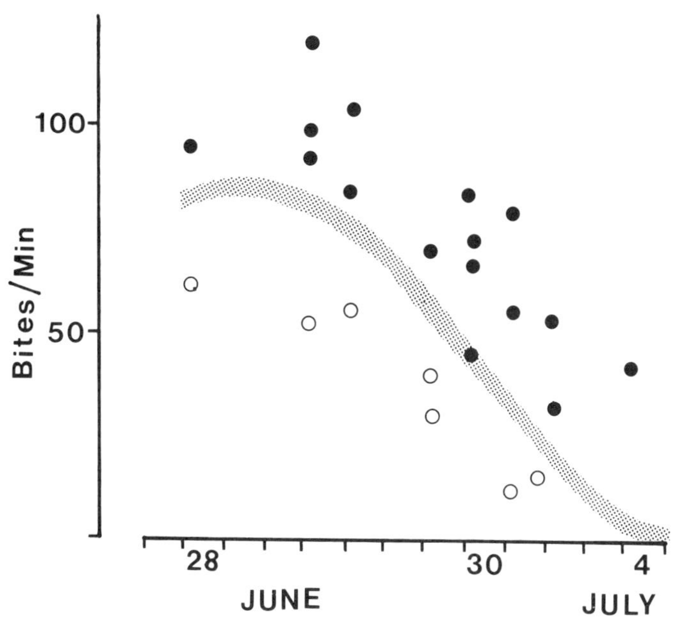

Figure 2.7 The intake rate of Barnacle Geese feeding on willow buds on patches of tundra in Spitsbergen. The birds are females during absences from the nest in the course of incubation. The mean intake rate at different times (the shaded band) was determined empirically. Plotted are the intake rates of individual birds which nested successfully (closed dots), and failed (open dots). Redrawn from Prop *et al.* (1984). There is a clear relationship between individual feeding performance and nesting success.

a new supply of willow buds, flowers and horsetail *Equisetum* is available as the snow-free patches increase in size. Prop *et al.* (1984) painstakingly counted or measured the food supply as birds emerged from the snow and then monitored the activity of individual, marked females as they visited these patches during incubation breaks. The feeding performance of individual females varied greatly depending on whether the same patch had been visited before. On one visit, a single female removed a third of the willow buds and nearly half the saxifrage flowers, so being first to the patch was extremely important. The feeding success of geese during incubation breaks was clearly related to their subsequent breeding performance (see also Chapter 3).

2.2.3 *Dabbling ducks*

The relationship between ducks and their food supplies are less well understood, mainly because they are more difficult to study. Early investigations involved analysing the contents of the stomach of ducks collected from shooters (e.g. Anderson 1959, Olney 1963). Whereas these studies established the main items of diet in the various species, they made only a minor contribution to the understanding of the way in which ducks interacted together and with the habitat and available food supplies.

Grazing ducks such as Wigeon behave very much like geese and similar methods have been used to study them. In places where there is little disturbance, they feed largely by day, staying close to the water's edge and radiating outwards to graze in short grasses or on leaves floating on the water surface (Owen and Thomas 1979). When disturbed they fly back to the water and start the process again. The importance of the safety of water has been demonstrated by Mayhew and Houston (1989), and illustrated in Figure 2.8. The birds are balancing the need to feed in areas where the biomass of vegetation is sufficient for successful foraging, and spending as little time as possible on the look-out for predators.

Most species of dabbling ducks are crepuscular or spend a proportion of the night feeding and are thus difficult to observe. Of those species that have been studied over the 24 hours, the herbivorous Wigeon (56% of the time feeding—Campredon 1981) and Gadwall (64%—Paulus 1984) need to spend substantial periods feeding. By contrast the seed-eaters and omnivores such as Mallard (35% of the time—Jorde *et al.* 1984) and Teal (42%—Tamisier 1972) manage on shorter feeding bouts. When using rich food supplies such as waste grain from stubbles dabbling ducks fill their crops in an hour or two around dawn and dusk and spend most of the day loafing on the roost (Thomas 1981).

Several species feed mainly at night and there may be advantages in terms of protection from diurnally active predators, such as has been demonstrated for Teal in the Camargue, France (Tamisier 1974). The heat generated by the feeding activity and digestion (**specific dynamic action**) also makes it advantageous for ducks to remain inactive during the day and, if possible bask in the sun, and feed during the relatively colder night (Jorde and Owen 1988). For intertidal estuarine feeders like the Shelduck, the timing of activity is necessarily dictated by the tidal rhythms (Bryant and Leng 1975). Although there are no data, we assume that feeding proceeds at night as well as during the day.

The availability of foods to dabbling ducks, which can forage effectively only in marshy areas or in shallow water is affected by ground conditions as well as the actual standing stock of the food. Having found suitable conditions, since they cannot see submerged seeds or invertebrates, the birds presumably go through a process of sampling to find the areas where

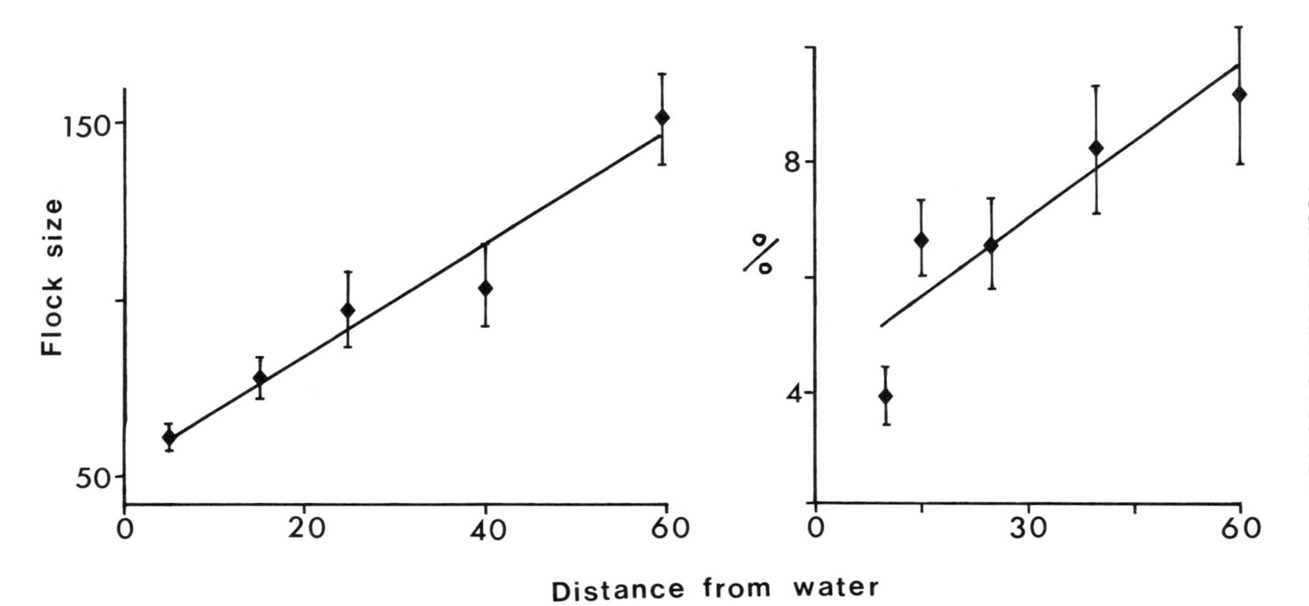

Figure 2.8 The regression of flock size (left correlation coefficient $r = 0.535$, $P < 0.001$) and the percentage of time spent vigilant (right $r = 0.344$, $P < 0.001$) on the distance of Wigeon from water. From Mayhew and Houston (1989).

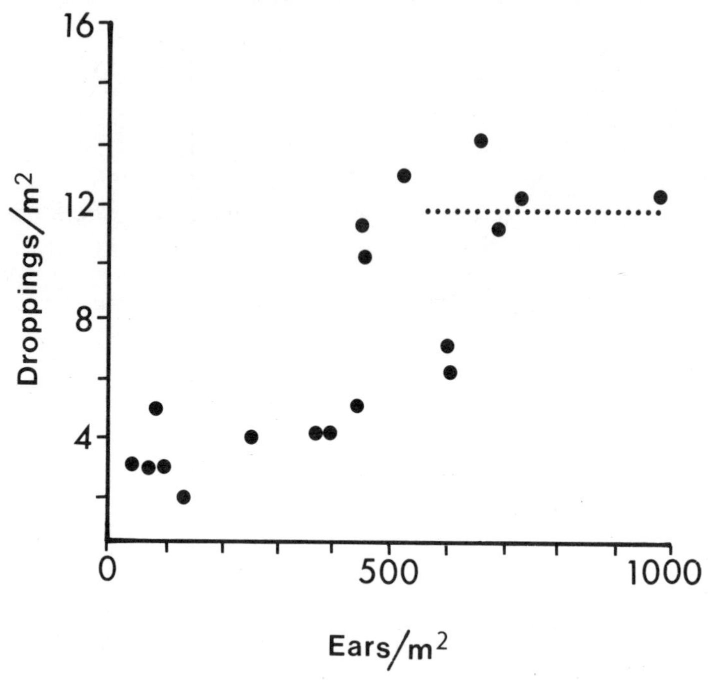

Figure 2.9 The grazing intensity by Teal, as measured by the density of droppings on an intertidal area in the Netherlands, in relation to the stocks of their chief food (density of *Agrostis* ears per square metre). From Van Eerden (1984).

food density is highest. Teal feeding on (*Agrostis*) grass in a Netherlands saline habitat matched the intensity of feeding (which was all at night) to the density of seeds (Figure 2.9). Van Eerden (1984) suggests, by the dotted line, fitted by eye to the points, that there is a threshold level (around 400 ears/m^2) below which the ducks cannot obtain sufficient food to merit the effort. The fact that there are droppings in areas of much lower densities of food than this indicates that the birds are sampling the whole area, but concentrating their feeding on the most profitable.

2.2.4 *Diving ducks*

Diving ducks capitalise on various underwater foods. There is very little information on the detailed activities of most of the species since they are often nocturnal and difficult to approach. The Canvasback of North

America is almost entirely vegetarian in winter; it subsists largely on the storage organs of underwater plants. Indeed, its specific name *vallisneria* comes from that of the wild celery *Vallisneria americana*, the wintering buds of which is the Canvasback's principal food. These storage organs are high in carbohydrates which supply the high energy demands of migration and wintering. In one Mississippi River study area diving ducks, mainly Canvasbacks, removed as much as 40% of the standing stock of wintering buds in a period of only a few weeks on autumn migration (Korschgen *et al.* 1988).

Most species of *Aythya* ducks can be seen resting on the water during the day; most of their feeding activity takes place often in different locations from the resting flocks. Tufted Ducks, which were feeding largely on the Zebra Mussel *Dreissenia polymorpha*, on Swiss lakes fed almost entirely at night and spent less than 5 of the 24 hours (21%) diving (Pedroli 1982). The mussel is a concentrated food source and rather easily gathered; in situations where the ducks are feeding by sieving the bottom mud the species can be expected to need a greater proportion of its time feeding. The fact that diving ducks are active preferentially at night and that they can gather their daily needs solely in the hours of darkness means that they can coexist with water-based recreation provided there are undisturbed daytime roosts available.

2.2.5 *Sawbills*

During the breeding season on a river in Sweden, both Goosanders and Red-breasted Mergansers foraged primarily by day. Their activity was well correlated with that of their main prey—the River Lamprey *Lampetra fluviatilis*. Later in the season the Goosander switched to feeding in the evening, following the main activity periods of the spawning Lamprey. The merganser remained diurnal, concentrating on the Three-spined Stickleback *Gasterosteus aculeatus* which is active by day (Sjoberg 1985). Both species hunt primarily by sight, initially searching the bottom for fish by immersing only the head, so it makes sense for them to be active by day. They can, however, locate fish in turbid waters by probing in crevices between stones (Sjoberg 1988).

2.3 Food selection

All other things being equal, animals are generally found to distribute themselves in the habitat according to the quantity and quality of food

(Fretwell 1972). Geese are exclusively vegetarian and have, in comparison for example with grazing mammals, rather poor digestive systems and low digestive efficiency. The availability of food to Brent Geese is to some extent governed by disturbance but their exploitation of food through the winter has been shown to be directly related to the quality of that food. Drent *et al.* (1980) suggested that the energy assimilated by the birds from particular foods would prove to be the best indication of its value. When Brent arrive on the wintering grounds, they graze preferentially on *Zostera* beds. As the *Zostera* is depleted later in the season by the grazing of the geese and by natural die-back, the birds turn to feeding on the green alga *Enteromorpha*, followed by saltmarsh grasses and growing cereals from inland fields (Charman 1980). Both Charman and Drent *et al.* found that the birds were obtaining more nutrients per unit time from *Zostera*, followed by *Enteromorpha* and grasses. That is, the birds were moving down a profitability gradient and their main criterion of profitability was the energy gain per unit time.

The subject of the effect of repellent or toxic products on food selection by wild waterfowl has been largely neglected, but plants have a number of defences against predation by animals. Fibre is not only virtually indigestible itself, it has a depressing effect on the digestibility of other nutrients (see e.g. Drent *et al.* 1980). Plant secondary products sometimes deter their grazing by herbivores. The content of phenolic compounds was shown to be an important criterion of food selection in the wild and in feeding experiments with Canada Geese (Buchsbaum *et al.* 1984). The addition of phenolic extract significally depressed selection of palatable species. In that study the presence of inhibiting compounds was a better predictor of selection than nutrient content. The role of gross or digestible nutrients or energy in food selection has been clearly demonstrated in many situations, however, but the importance of inhibitors deserves more study.

In one of the few studies of food selection and profitability in diving ducks, Draulans (1982), designed a series of experiments to test whether Tufted Ducks, when given a choice of mussels of different sizes, chose the most profitable. Different sized mussels took differing amounts of time to deal with (handling time); the relationship and the profitability of mussels of different sizes is shown in Figure 2.10.

The captive experiments were supported by work on the natural situation. Draulans manipulated the density of mussels by scuba-diving and first denuding an area of mussels and then replacing them at known densities. The mussels that remained after natural predation by the ducks

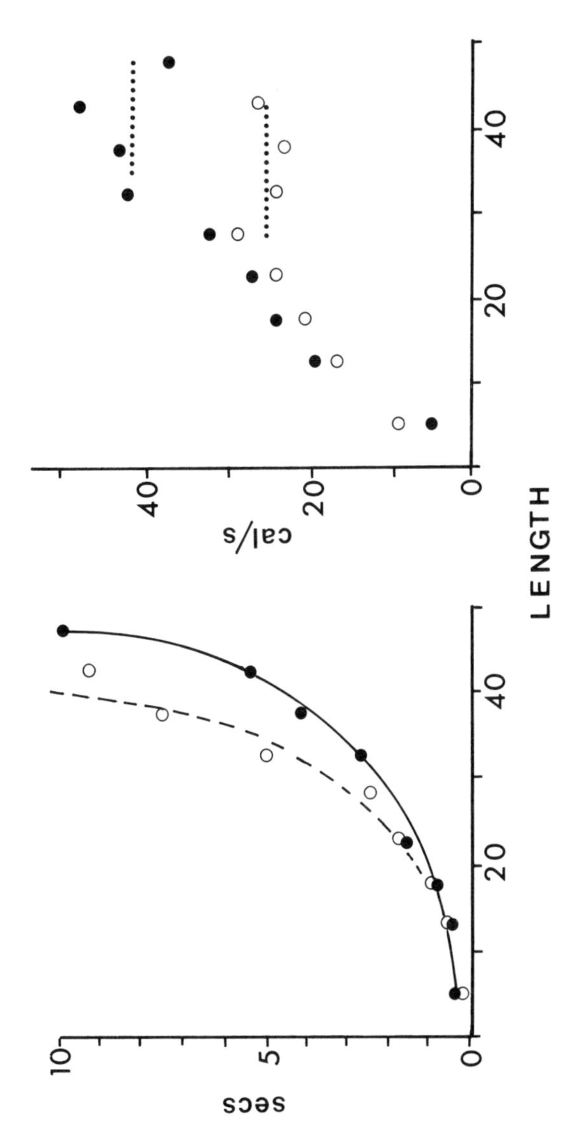

Figure 2.10 The handling times of mussels in relation to their sizes (left) and their profitability (energy gain per unit time) to feeding Tufted Ducks in Switzerland. Two ducks were involved in the trials: closed dots—a male and open dots—a female. After Draulans (1982).

were collected and their sizes compared with those of all the mussels presented.

The experiments in captivity, which used a single pair of ducks, indicated that they were highly selective in the sizes of mussels they ate, and that these were similar to sizes taken under natural conditions. The experiments in the wild showed that Tufted Ducks concentrated their feeding in the areas of highest prey density. They also indicated that different sizes were selected when different amounts were available. At very high densities the birds took smaller and rather more variable sizes than at low density.

Contrary to expectation, the birds selected mussels that were smaller than the most profitable in terms of energy input per unit time. There were several possible reasons for the discrepancy. Under natural conditions, when the time spent at the bottom searching and handling food was limited, the ducks could eat two small mussels rather than one large one. At high prey density the ducks did not have to search for the mussels and so could eat two small ones making the choice of small mussels the more profitable option. At lower densities, however, the added searching time meant that only one mussel could be consumed and it was more profitable to take a large one. Draulans also thought that the ducks may have avoided large mussels in case they selected one which was too large to swallow.

Because sawbills pose a threat to salmonid fisheries, whether commercial or sporting, their feeding behaviour, food selection and consumption have been given a great deal of attention by scientists. Fish-eating mergansers pursue very active prey and for them the capability and activity of the prey as well as the predator's hunting skills are important.

Food selection by sawbills has also been tested in detail in experiments in captivity. Different species of fish reacted differently to the presence of the predator, those inhabiting open water tried to escape, whilst bottom dwellers tried to hide among stones. This made them differentially vulnerable to predation. When presented with fish in a situation where the prey had little chance of escape, the birds showed a preference, but this was not reflected in the prey caught under semi-natural conditions in a stream tank, indicating that prey behaviour is important. Rather, preference was related to the 'handling' time of the prey (Sjoberg 1988). The predators operated in general according to profitability rules, modified by the behaviour of the prey.

In a more natural situation, in an enclosed natural river in Canada, Goosanders were, as might be expected, more successful at higher density of prey. They were also more successful in stretches of river where the fish had fewer places to hide. After fish had experienced a Goosander fishing, they were less vulnerable because they were more alert to danger and more likely to retreat to cover (Wood and Hand 1985).

2.4 The annual energy cycle

This chapter has described in brief the way in which waterfowl exploit the wide variety of aquatic habitats and foods found there. Most of the information deals with the wintering area; since we have argued that it is there that the adaptive radiation of structure and behaviour has developed, we feel justified in giving the breeding grounds only a cursory treatment. But most waterfowl are migratory or nomadic and each species has to fit its nutrient and energy requirements not only to its present, but also to some extent future needs. This section gives examples for well-studied species, of the way in which the birds fulfil the different demands of the various stages in the life cycle.

2.4.1 *The Lesser Snow Goose*

Geese and swans are very closely related and have very similar life history strategies. Nearly all species are migratory to some extent and most have long migrations from high to low latitudes. All species are vegetarian, though swans are aquatic and are nearly always associated with water, whereas geese do much of their feeding on dry land. In terms of the annual cycle, however, there are rather few differences between the migratory species, so we will use a goose species as an example for the Anserinae. The seasonal variation in body weights of male and female Snow Geese are shown in Figure 2.11 and this can be regarded as typical for migratory species.

Autumn is generally a time of plenty; geese and swans switch from grazing to digging or seed feeding. Perennial plants draw nutrient from leaves and stems to storage organs, usually at or near ground level. These include roots, tubers, stolons, bulbs and other structures which store nutrients and energy, usually in the form of starch. These energy stores not only carry the plant through the winter but also enable growth before the temperature has risen sufficiently to allow growth from assimilated nutrients in the spring. These energy-rich foods are also easily digested and provide a very valuable item of diet for waterfowl. Seeds are also rich in energy and many species of geese and swans strip seeds from standing stalks or sieve them out by dabbling in shallow water.

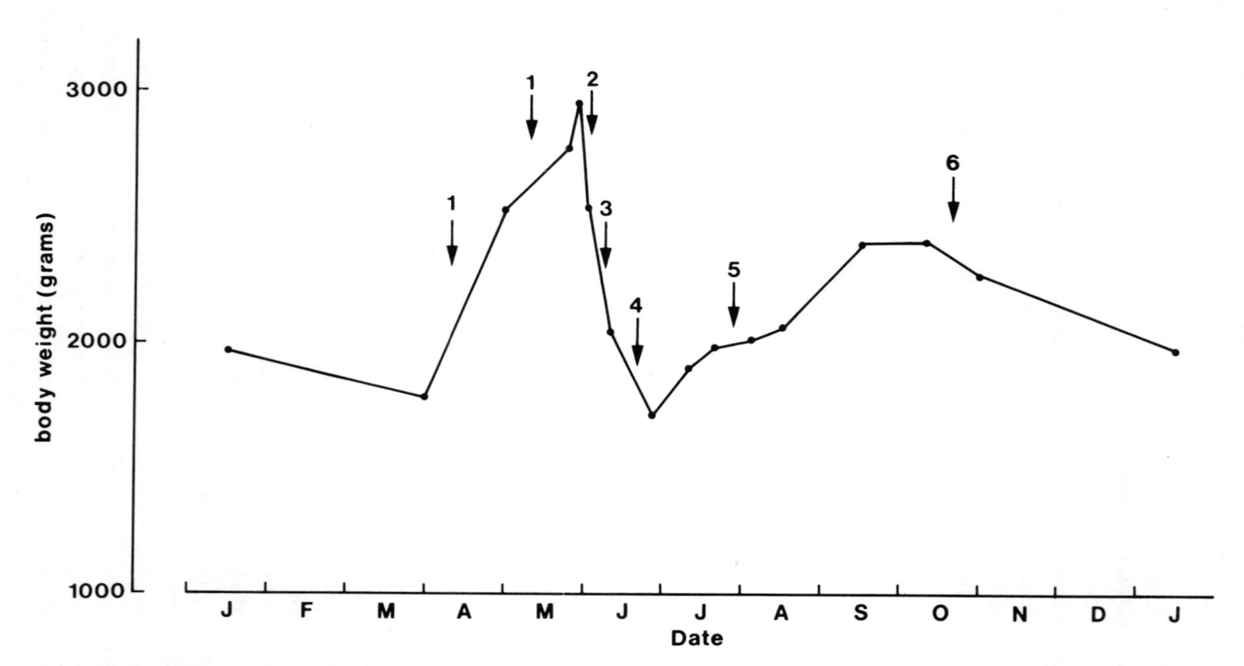

Figure 2.11 The pattern of weight change in Lesser Snow Goose through the annual cycle. The vertical arrows indicate the following events: 1—spring migration; 2—laying; 3—early incubation; 4—late incubation; 5—moult; 6—autumn migration. Based on a number of studies in North America. From Owen (1980a), sources quoted there.

The daylength in September and October is long enough to allow extended periods of feeding; all species gain weight at this time. This is partly a recovery after the stresses of migration and breeding, and replenishment of nutrients depleted in summer, but also fat reserves are being laid down for use, if necessary, in harsh weather. Fattening in the Snow Goose starts in the autumn staging areas on James Bay, Canada; as they leave for the south adult males weigh nearly 2.7 kg and a high proportion of this is made up of fat (Wypkema and Ankney 1979).

In early winter the birds maintain their weight but by January the effect of short days and cold weather have necessitated some of the reserves to be used up when food is limiting. In midwinter male Snow Geese weigh just over 2 kg (Flickinger and Bolen 1979). There is some debate over whether the loss of reserves is due to stress, or whether the birds store them as a precaution against severe weather and use them up even if those conditions do not occur. There is evidence from wading birds (reviewed by Pienkowski *et al.* 1984) that the latter is more likely, and this is probably true of geese and swans. In general, birds maintain only sufficient reserves for their (anticipated) needs rather than carry extra weight at the cost of losing manoeuvrability in flight.

In the spring, rising temperatures promote the growth of vegetation and the first flush is high in protein and easily digested. The days are lengthening too, and the birds begin to lay down substantial reserves in preparation for migration and breeding. Migration is usually in several stages (see Chapter 5); since movement is up a latitudinal or altitudinal gradient the birds are effectively capitalising on a series of springs, thus extending the period when they can capitalise on vegetation in a highly nutritious and digestible stage. This has been vividly described as 'riding the crests of digestibility waves' (Drent *et al.* 1980). On arrival on the breeding grounds, Snow Geese are at their highest weight of the cycle; males exceed 2.8 kg. Significantly this is the only time when the female is heavier than her mate; her energy requirements for nesting and incubation are much greater than those of the male (Ankney and MacInnes 1978).

Geese and swans arrive on the breeding grounds when there is little or no food available, so the nutrients for the eggs and the energy to sustain the female during incubation have to be provided largely from reserves. The amount of feeding that is possible varies with the species; in general the higher the breeding latitude the less is the opportunity for feeding during incubation. The weight of the female Snow Goose at the end of incubation is reduced to 1.7 kg—the lowest of the life cycle, and only just over half the peak weight only a month previously. The male retains condition during incubation and largely takes over parental duties during the moult, when the female gains weight in preparation for autumn migration. The period of the flightless moult is one of high metabolic demand, but the available evidence suggests that this is not necessarily a period of great stress (Ankney 1979). The birds prepare by laying down some reserves beforehand and compensate for the high energy demands of feather growth by being relatively inactive (Owen and Ogilvie 1979). As soon as the birds regain their powers of flight they have access to rich feeding areas and the autumn fattening period begins again. The ability of the birds to be able to complete the autumn migration successfully crucially depends on their gaining sufficient reserves to fuel the flight (see page 120).

As well as energy and nutrients, the birds also use during breeding, reserves of elements which may be limiting in the diet. Female Snow Geese transfer calcium from the bones into eggshells during laying and this is replaced later (Ankney and MacInnes 1978). The moult requires high levels of sulphur-rich amino acids and although evidence is lacking these may also be supplemented from reserves.

2.4.2 *The Mallard*

Dabbling ducks take a variety of foods, but in winter most species either sieve seeds and invertebrates from shallow water or browse on vegetative parts of plants. As with geese and swans, there is an increase in body weight in the autumn, in preparation for the winter. The variation in the body condition of Mallard in southern England over nine months and in central North America for the other three months are shown in Figure 2.12. Females, which have to rear the brood, and to delay their moult, reach their peak condition later than males. After December, there is a drop in condition, and this is related to a switch from grain feeding to natural foods rather than to cold weather (Owen and Cook 1977). Although Mallard in their study used farmland for feeding these weight patterns probably reflect those in more natural situations. There is a loss of condition related to severe weather, but in the coldest winters, the birds move south or to the coast.

The expected increase in condition in spring starts as early as March in the Mallard, which is an early nester (Figure 2.12). It is difficult to say whether the rapid increase in weight between March and April represents the real scale of increase since the figures are from different studies, but the species is comparable in size on the different continents. The ducks are at

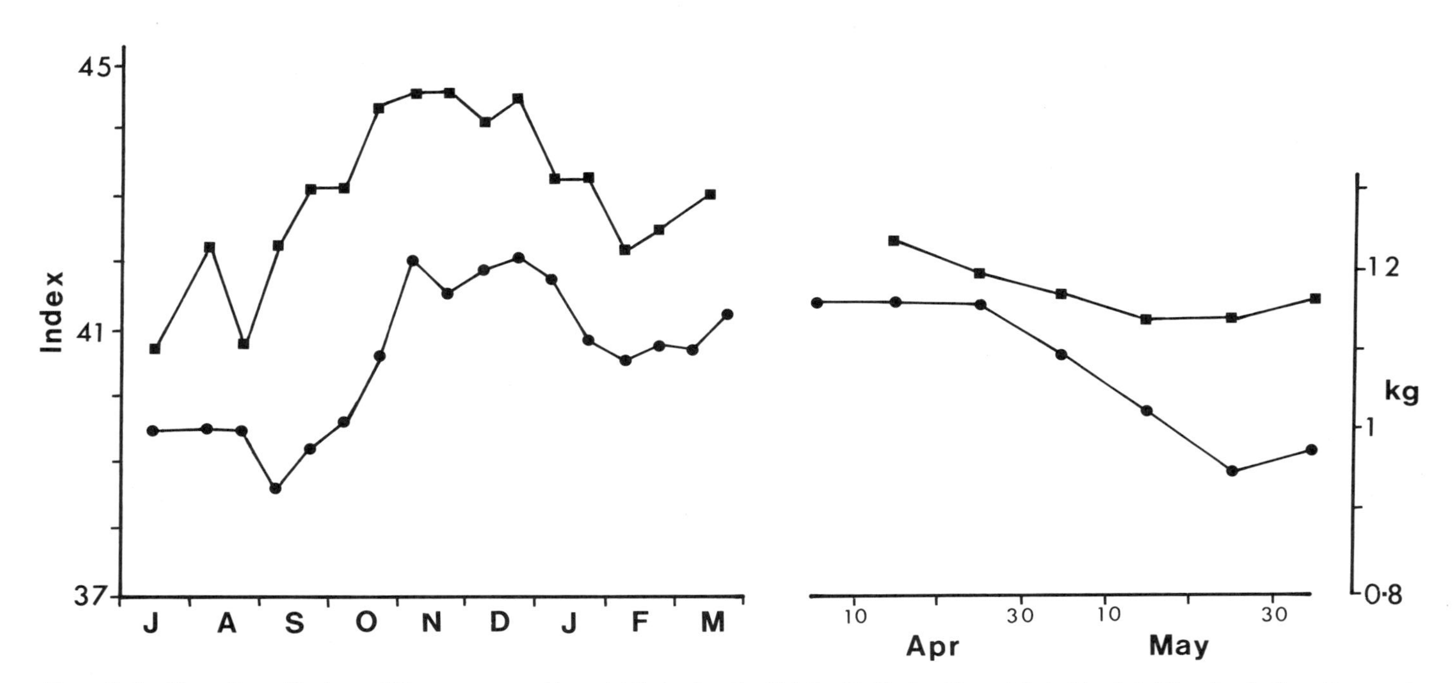

Figure 2.12 The pattern of body condition, as expressed by weight/wing length of Mallard in England from July to March (left hand scale, from Owen and Cook 1977), and of the body weights of Mallard in North Dakota during the breeding season (right hand scale, from Krapu 1981). Squares—male and dots—females.

their heaviest on arrival on the breeding grounds; the males lose weight in the early days, when they are defending their mates and feeding areas for them. Females draw on their body reserves to provide energy during laying and incubation; during the nesting cycle the female loses about 25% of her weight (Krapu 1981). The protein needed for egg prodution is, however, obtained from feeding, as the ducks switch to feeding on a chiefly invertebrate diet, as do other dabbling ducks at this time (e.g. Krapu 1974). Since body reserves carried to the breeding grounds make a substantial contribution to the breeding effort, at least of early nesters, we might expect that there is a relationship between winter conditions and breeding success. These is little evidence on this point, but the positive relationship between precipitation in winter and the proportion of young ducks in the following autumn (Heitmeyer and Fredrikson 1981) is suggestive. Because the conditions on the breeding ground can mask completely variations in success due to differing body condition of arriving birds, the relationship between wintering conditions and dabbling duck breeding success, if it exists, will remain difficult to demonstrate.

The loss of weight in male Mallard in late August and in females in early September is coincident with the flightless moult. As in geese, the weight loss is pre-programmed rather than being the result of stress. The fact that Mallard moulting in enclosures of the Wildfowl and Wetlands Trust, where ample supplementary food is supplied, nevertheless lose weight during moult, adds support to this.

2.4.3 *The Ring-necked Duck*

The Ring-necked Duck is a small diving duck which specialises in rather unproductive bog and marsh habitats in North America. Its pattern of energy demand follows that of Mallard. Both males and females rely heavily on nutrient reserves laid down before breeding. There is a recovery immediately following nesting and a loss of weight during the moult. The autumn, when the Ring-necked Duck's diet consists mainly of seeds, is a time of fattening, and there is a decline in midwinter (Hohner *et al.* 1988).

2.4.4 *The Eider Duck*

The Eider Duck is an extreme case among ducks in that the female relies entirely on her body reserves for both laying and incubation (Milne 1976, Korschgen 1977). At the end of incubation the female weighs as little as half her prelaying weight. Fat and protein for egg production are obtained from body reserves despite the fact that food is available close to nesting islands. This is almost certainly an adaptation to the very high levels of potential predation of eggs (Milne 1976). Gulls and skuas are voracious predators of unattended nests; female Eiders cover their eggs during laying and leave the eggs for very short periods to drink and bathe during incubation.

2.5 Summary

Waterfowl show a wide variety of feeding behaviours, related to the activity of predators and to the availability and activity of their prey. They select food in relation to its profitability, which can usually be simplified to the amount of energy (or at some seasons the amount of specific nutrients) gained per unit time. They can sometimes manipulate the growth of plants to maintain its quality and in some cases in breeding areas may have a great impact on the vegetation. They match their needs according to the present and future demands of the life cycle, storing reserves in the body in times of plenty and using them up for energetically demanding activities such as breeding or migration.

CHAPTER THREE
BREEDING BIOLOGY

This chapter deals with breeding ecology of waterfowl on a species or individual level. The factors controlling reproductive performance and recruitment into populations are covered in Chapter 6.

3.1 Timing of pair formation

Waterfowl are unusual among birds in that pair formation takes place in winter, some months before the breeding season. Most workers have linked early pairing to the timing of the breeding season and suggested that the defence of the female by the male during the pre-laying and laying period confers advantages in that nutrients for maintenance during laying and incubation can be stored some weeks in advance (see e.g. Lamprecht 1989, also Chapter 2). Rohwer and Anderson (1988) undertook a detailed review of this and other unusual aspects of waterfowl biology and we summarise and assess their conclusions below. They examine several alternative hypotheses to account for early pairing; these had been proposed by a number of workers and evidence to test them was gathered from a variety of sources, which we do not quote here. Readers are referred to their review for details. The following hypotheses were examined.

Time of breeding. Rohwer and Anderson found little conclusive evidence that the timing of pair formation was determined by the earliness of nesting; indeed they found that between species, there was no correlation between the earliness of pairing and the timing of breeding either in dabbling ducks or diving ducks. They presented no evidence linking the timing of nesting or its success with pairing date within a species. Owen *et al.* (1988) did present such evidence for Barnacle Geese, and suggested that the time available for fattening and pair familiarisation significantly influenced breeding success (see page 70).

Mate testing. This hypothesis suggests that birds that pair early will sample more potential mates than those that do so later. Trial periods or liaisons frequently occur in ducks (McKinney 1986) and geese (Black and Owen 1988). During these liaisons partners take part in mutual social display which often involves aggression towards neighbours. However, there is no information from within or between species comparisons that this behaviour is linked with either the duration of the pair formation period or the timing of breeding.

Diet quality. Paired waterfowl have a higher rank than unpaired ones and are able to gain access to the best food. There is a conflict between the idea that birds under the greatest pressure for feeding would on the one hand not be able to spend time on courtship and pairing but on the other would have most to gain in a competitive situation from the elevated social status that being paired confers. Rohwer and Anderson assume that males as well as females must benefit directly for early pairing to be beneficial according to this hypothesis. However, a male benefits automatically through the breeding success of his mate provided his paternity is guaranteed. There is certainly evidence that being paired in the pre-migratory and pre-breeding fattening period is advantageous to a number of species, since it allows the female, whose breeding demands are greater than those of her mate, to lay down greater reserves (Ashcroft 1976, Teunissen *et al.* 1985).

Male/male competition. This assumes that pair maintenance is costly to a male, in that his investment in energy is proportional to the time. On the other hand early pairing birds have access to a greater pool of mates, so they are more likely to pair with the best females, a pressure for early pairing. Rohwer and Anderson tested a prediction of this hypothesis—that pairing would be earlier in species with a more male-biased sex ratio. There was no correlation in either *Anas* or *Aythya*, so the hypothesis was rejected.

Accumulation of nutrients. Effectively this is not distinguishable from the hypothesis involving diet quality, since food quality in most species during the pre-breeding period means greater nutrient accumulation. Rohwer and Anderson present little evidence in support.

Male costs and female benefits. This is an all-inclusive hypothesis suggesting a balance of costs for males (increased energy demands and greater mortality) and benefits to females (feeding advantages). This is Rohwer and Anderson's preferred hypothesis. They argue that males in good condition can afford to pair early and that larger species are at an advantage because of this. They provided very weak evidence in support, however.

The benefits to females are derived from the better feeding opportunities and greater nutrient accumulation with a longer paired period. We agree with this suggestion, but the generally later pairing of diving versus dabbling ducks is still unexplained. Rohwer and Anderson argue that the explanation that males are less able to defend the (submerged) food supply than are surface feeders explains this. However, one of the clearest advantages of pairing in ducks is for a diving species (Ashcroft 1976).

Rohwer and Anderson present a series of speculations and suggestions in support of this final hypothesis, but they are not convincing, especially on the costs to either sex of being paired or the benefits to males of the eventual breeding success of his mate. The potential benefit to a male of increasing its productivity overwhelms the costs of pair maintenance. We suggest, therefore, that the benefits for both partners arises through nutrient acquisition for the female. Variation within and between species are likely to be imposed by the competing demands of food gathering, as it constrains the time available for courtship and pair formation. For example, Brodsky and Weatherhead (1985) found that courtship in Black Ducks started earlier in a site with a good food supply as opposed to a poor feeding area. Lamprecht (1989) concluded, having examined three alternative hypotheses, that the support of female feeding provided the best explanation of pair proximity in the Bar-headed Goose.

3.2 The timing of nesting

The ultimate factor controlling the timing of nesting is that which controls the number of young produced—usually in birds the amount of food available for the growing chicks. The proximate factors—those that trigger off the breeding cycle—have to operate some time beforehand so that the young hatch at the time of maximal food production. There may be various constraints which modify a bird's ability to match its reproductive efforts to the environment. Among these may be the unavailability of nest sites because of snow cover or shortage of food for the laying female. These constraints may operate at the individual or the population level.

The time between the possibility of breeding (when, for example, nest sites become available), together with the time required for laying and incubation, and the peak abundance of food may be too short for a bird to complete laying and incubation before the 'best' time for hatching. In that case the young might hatch at a suboptimal time. For example, in a northern Swedish lake, ducks began nesting as soon as the thaw set in, before the chironomid flies begin to emerge. The peak insect emergence time occurs about a month after the thaw and ducklings hatch soon afterwards. Hatching almost coincides with the period of highest food abundance (Danell and Sjoberg 1977). Similarly, Cackling Geese in Alaska

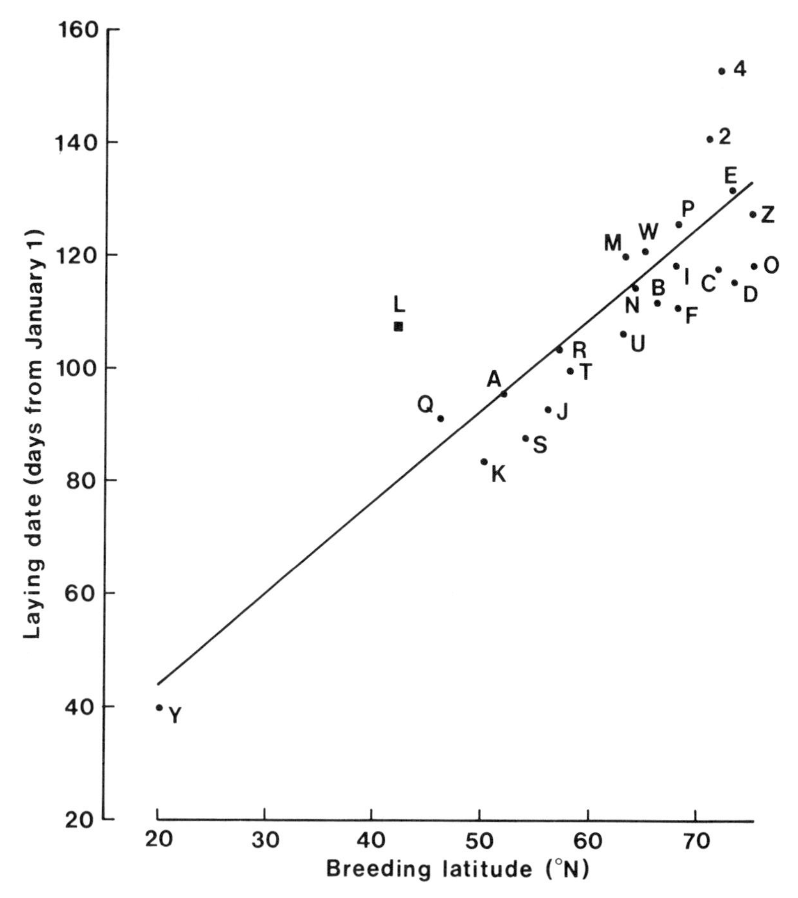

Figure 3.1 The relationship between the mean date of laying of the first egg at Slimbridge and the mid-latitude of the breeding range in the wild for Geese. Updated version of that in Murton and Kear (1973). The correlation coefficient, excepting the Bar-headed Goose (L), which breeds at high altitudes but at relatively low latitude $r = 0.913$, $P < 0.001$. Reproduced from Owen (1980a). For explanation of letters see Appendix.

normally hatch after the peak in the nitrogen content of the vegetation (Sedinger and Raveling 1986). In most species, there is a range of variation between individuals in the timing of laying so that only some birds breed at the 'best' time, although there is a high degree of synchrony in some species as discussed later.

Except at very low latitudes the cycle of daylength is the most predictable annual variable that triggers off the breeding cycle. The effect of daylength in the family Anatidae has been examined in captivity in a kind of accidental 'experiment' where birds breeding at various latitudes have been transported to a single location and the timing of their breeding season recorded.

The median laying dates of each species is related to the latitude of its native range—a clear relationship with daylength. An example, that of the true geese, is shown in Figure 3.1. The length of day required to trigger off nesting increases with latitude; the most southerly species, the Hawaiian Goose, breeds at Slimbridge, England, when the daylength is only 9–10 hours, whereas the most northerly geese do not start to lay until there are at least 14 hours of daylight. The Hawaiian Goose does not nest after the length of the day exceeds 13.8 hours, suggesting that egg formation is inhibited by long days (Murton and Kear 1973). Two species seem not to conform closely to the relationship in Figure 3.1; the Bar-headed Goose (L), nests at 42° but at a daylength similar to those breeding at 60°. This species nests at very high altitudes in the Himalayas, which are, climatologically, similar to locations at much higher latitudes. The mid-latitude of the breeding range of the Red-breasted Goose (4) is at 70° North, though its laying date is typical of species breeding at much higher latitudes, i.e. it lays much later than expected. This species is exceptional among geese in that it apparently nests exclusively in association with birds of prey, and it has been suggested that it delays laying in order to synchronise with the nesting time of its associates (see Owen 1980a, p. 111). The relationship between laying date at Slimbridge and latitude of origin is consistent for all non-tropical waterfowl, providing strong evidence that daylength is a strong proximate factor initiating breeding in most species (Murton and Kear 1973).

The variability of laying date within a species depends on the stringency of the breeding conditions. Among the geese, the low-latitude breeders such as the Nene and the Greylag, show considerable variation in laying date at Slimbridge, whereas range of dates for the high-latitude breeders is very narrow. In the wild, selection for laying at the right time in northerly breeders is severe; the 'window' of opportunity for nesting successfully is small and this reduces the genetic variation in the populations.

In the tropics, there is very little variation in daylength but the occurrence of rain is usually seasonal, so that suitable breeding habitat becomes available on a roughly annual basis. In inland Australia rains are not very predictable, and ducks breed erratically when conditions are suitable. Figure 3.2 shows diagrammatically the relationship between water levels in inland billabongs (temporary river floods) and the timing of breeding in four species of ducks. As the water level rises following heavy rains new feeding areas become available and the Grey Teal responds very quickly, laying eggs within a few days of the start of rising water. The Black Duck follows but the Australian White-eye or Hardhead, a diving duck, delays its nesting until there is sufficient deep water for feeding. The Pink-eared Duck is a plankton feeder and its food is most abundant as the water level recedes; it is the latest of the four species to lay. The way in which the breeding cycle is controlled in these species is not fully understood, but the Black Duck, and probably the Hardhead, do have an annual cycle of gonadal activity which is related to daylength, whereas the Grey Teal and the Pink-eared Duck show no photoperiodic breeding response (Braithwaite 1976, Frith 1982). The last two species are able to breed at any time of year, whenever ecological conditions are suitable.

Whatever the latitude, it is the ability of an individual to time its breeding effort as closely as possible to the optimum time that is the main determinant of its success; individuals vary in their genetic make-up and in their experience and these, tested in competition with other individuals of the same and different species determine individual productivity.

3.3 Breeding range

Competition between species and their different adaptations to environmental conditions usually result in separate breeding ranges or habitats. However, if, as we argue in Chapter 6, it is in winter that population sizes are determined, although there may be considerable overlap between species in habitats and foods in summer, there may be little competition. In Arctic geese and swans although there are some variations in growth rates, the fledging period is closely related to adult body size, which is the main determinant of the northerly limit of the breeding range (Owen 1980a).

The determinants of the southerly limit are less clear, but Owen (1980a) suggested that this was set by the protein density in plants which determines the speed of growth. However, geese of widely different adaptations nest together in areas such as the Yukon–Kuskokwim Delta in Alaska, and a colony of Barnacle Geese has recently been established in the Baltic, far to

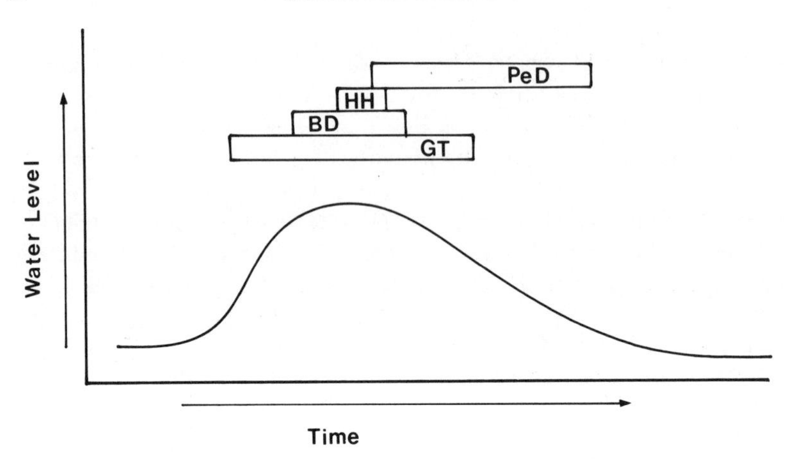

Figure 3.2 Diagrammatic representation of the relationship between the breeding seasons of four species of Australian ducks and changes in water levels at inland billabongs. BD Black Duck, HH Hardhead, GT Grey Teal and PeD Pink-earned Duck. Redrawn from Frith (1982).

the south of the native range of the species (Larsson *et al.* 1988). The habitat used by the Baltic population has been modified by Man, so their situation is not natural. However, the nutritional hypothesis does not entirely explain the southerly distribution. Perhaps the birds naturally nested at the most northerly possible latitude for nutritional reasons, and that the number of birds in the population (regulated on the wintering areas), determined how far south the range had to extend (see Gauthreaux 1982 for further discussion of this subject).

The factors affecting the distribution of ducks are clearly related to their needs for nesting sites and foods. Wood and Mandarin Ducks nest and readily perch in trees and their favoured foods are acorns, which restricts them to southern deciduous forests. The hole-nesting Goldeneyes breed in the coniferous forests further north. Many species do coexist in such places as Myvatn in Iceland and the prairie pothole region of North America, where food in normal seasons is superabundant. If breeding habitat were equally available, we would expect that the breeding areas closest to the wintering range would be colonised first, in order to minimise the length of hazardous migratory journeys. This does seem to be the case, at least for some ducks (Nudds 1978), although the evidence is somewhat circumstantial. Johnson and Grier (1988) also found the highest breeding densities at the lowest latitudes in four species of North American ducks. In others, however, they argued that the most suitable habitat for breeding was not that closest to the wintering areas and that species such as the Blue-winged Teal and Redhead had evolved homing patterns which meant that they overflew sub-optimal habitats so that it was the most suitable areas that filled up first. During exceptional drought years in the prairies, there is a northerly shift in the optimal habitat and the ducks continue their migration northwards (Hansen and McKnight 1964).

3.4 Nest sites

The most important characteristic of a nest site is its safety from predators; the egg stage is the most vulnerable in the reproductive process (see Chapter 6). The nest locations and characteristics of the whole group are given in Table 3.1. Northern swans are vulnerable only to land predators, and take no steps to camouflage the nest, which is usually on an island or on a floating platform of reeds. Islands are favoured locations for a large number of species and support very high densities of nests. Although Ross' Geese are aggressive and territorial, they pack closely on lake islands (Ryder 1967). Similarly, a 42 ha island in Loch Leven, Scotland, supports over 1000 nests of ducks, of six different species (Newton and Campbell 1975).

Table 3.1 The nest sites and construction in waterfowl groups. From Kear (1970).

	Over water	On ground	Burrow	Tree holes	Parasite	Sex building
Magpie Goose	*					M + F
Whistling Ducks	*	*		*		M + F
White-backed Duck	*					M + F
Cygnus Swans	*	*				M + F
Coscoroba Swan	*					M
Cape Barren Goose		*				M
True geese	*	*				F
Freckled Duck	*	*				?
Steamer Duck		*				F
Sheldgeese		*				F
Shelducks		*	*	*		F
Cairinini				*		F
Anatini		*		*		F
Aythyini	*	*			*	F
Mergini		*		*		F
Oxyurini	*				*	F

Cliffs are favoured nesting sites for several species of *Branta* geese, especially the smaller species which are unable to defend their nests against foxes. Open tundra breeders place their nests on outcrops or against rocks, which makes them more easily defended. Nests are sometimes associated with those of predatory birds, which give them protection from foxes; nests of Red-breasted Geese are almost invariably associated with those of Peregrine Falcons *Falco peregrinus* or Rough-legged Buzzards *Buteo lagopus*. Ducks, too nest in association with other birds, such as gulls and terns, which 'mob' intruding predators and keep them at bay. In the Loch Leven study, nests within a gullery were more successful than those outside and predation of duck nests decreased further towards the centre of the gull colony. The pressure from predators is so great that some ducks take advantage of artificial protection. A large colony of Eiders became established around a group of tethered huskies in Greenland (Meltofte 1978) and another is located at the end of the runway at Akureyri airport, Iceland. The eiderdown industry in Norway depends on the concentration of the breeding birds in houses specially constructed for them.

Other waterfowl seek protection in cavities among rocks or in trees. Hole-nesting species include Whistling Ducks, most of the perching duck tribe and the majority of the goldeneyes and mergansers. Several species, notably the Carolina or Wood Duck and its congener the Mandarin Duck, nest exclusively in tree cavities. As mature forests were felled hole nesters suffered but, luckily, most species take readily to nest boxes; the dramatic recovery of the Wood Duck as a result of the provision of artificial nests is described in Chapter 7.

Some ducks rely on other birds to provide nests for them. Egyptian and Spur-winged Geese sometimes use the old nests of birds such as herons, and the African Comb Duck almost always uses the nests of other species (Pitman 1965) and is closely associated with the Hammerkop *Scopus umbretta*, in whose abandoned large stick nests it breeds. The smaller goldeneyes and sawbills make use of old nests of woodpeckers; the distribution of the Smew in Eurasia matches closely that of the Black Woodpecker *Dryocopus martius* (Owen 1977a) and the Bufflehead in North America nests largely in the excavated holes of Flicker woodpeckers *Colaptes auratus* (Erskine 1972).

With a few exceptions waterfowl nests are simple structures—a slight excavation or scrape in the ground or cavity lined with a variable amount of down. Swan nests can be very large constructions, however, built up by the male and female (male only in the case of the Coscoroba Swan) of reeds and other material gathered from around the nest. The elevation is necessary insurance against rising water levels, which nevertheless account for some nest losses. Stifftails always nest over water and construct elevated nests of reeds or other emergent vegetation. The amount and function of down in waterfowl nests is dealt with in Section 3.7.

3.5 Breeding dispersion

Waterfowl exhibit a variable degree of territoriality, and species vary in the way in which they partition the habitat. Most swans are highly territorial throughout laying and incubation; only one pair can nest on small or medium sized ponds. Nests are normally out of sight of one another, but pairs tolerate one another if breeding conditions are particularly suitable, such as for Whooper Swans in parts of Iceland. In the highly territorial Mute Swan the size of the defended area is related to the quality of the habitat and the density of breeding birds. In the Oxford area, the average size of a nesting territory is 2.5–3 km of river, but territories are smaller on good habitats and larger on poor rivers and changes in density on the same stretches can be related to changes in the food supply (Bacon 1980). Some Mute Swans maintain territories throughout the year and these are also related to the food supply; the pairs on the best territories put most effort into defending them (Scott 1984). In places where the habitat is particularly suitable, Mute Swans even nest colonially (see e.g. Perrins and Ogilvie 1981).

The Black Swan nests colonially, even in areas where nesting habitat is superabundant. Breeding depends in many areas on the availability of temporary floods and when conditions favour breeding, food for the young is not limiting and there is no disadvantage to the colonial habit (Braithwaite 1982). Presumably colonial nesting is advantageous because of an anti-predator function.

Geese are generally territorial, though there are two distinct patterns of dispersion—colonial and dispersed nesting. The Snow Goose is the classic colonial breeder; pairs nest in close proximity even where there are large tracts of nesting habitat. A reproductive advantage of colonial breeding and synchronous hatching has been well demonstrated in the Lesser Snow Goose (Findlay and Cooke 1982). Geese nesting both earlier and later than the colony mean have poorer reproductive success than synchronous layers. This is because of lesser predation rates at peak laying and hatch, when the predators were 'swamped' by the sheer numbers of prey. Raveling (1989) demonstrated that the predation effects on Black Brant varied according to colony size. Predation rates in two small colonies varied between 55 and 85%, whereas in a large colony, losses were 31–32%.

Owen (1980a) suggested that there is a feeding advantage to colonial breeding, since the birds effectively increase the production and maintain the quality of the tundra vegetation (Harwood 1977). This is true in the early years of colony life, but as breeding density increased in one Canadian colony, Snow Geese overexploited the habitat, resulting in a food-related density-dependent decline in breeding success (Cooke 1990).

Many goose species are dispersed nesters; males defend areas around the nest during incubation from intruding birds but do not defend the female from predators. The nest is usually in some cover (Canada Geese, Bean Geese), or the female is very cryptically coloured and remains motionless when predators approach.

The ecological counterparts of true geese in the southern hemisphere — the South American sheldgeese—are also highly territorial; Kelp Goose males defend a section of shore where the family feeds. The female is cryptic whereas the male is brilliant white—its conspicuousness supposedly reinforces his territorial behaviour. Shelducks are also highly territorial; in the Common Shelduck the male guards a feeding territory to which the female flies during incubation and to which she brings the brood; the nest site itself is not guarded. The size of the territory of this species is very closely related to the food supply—the density of *Hydrobia* snails (Buxton 1975). This relationship is, however, not in the expected direction (territories are larger in better feeding areas). Patterson (1982) argues that territorial behaviour in the Shelduck limits the density of the population but does not relate numbers to the food supply.

Dabbling and diving ducks are variably territorial in the pre-laying and laying season, though it is a feeding area for the female that is defended rather than the nest site. The home range in most species is shared between several pairs, but chasing behaviour does lead to a greater dispersion of nests in species such as Shoveler (which is the most territorial of dabbling ducks) and Gadwall. It has been suggested that spacing has an advantage as an anti-predator device, making it difficult for predators to find nests scattered through the habitat (McKinney 1965).

Buffleheads nest in tree holes and are highly territorial during the nesting season. By providing a superabundance of nest boxes, Gauthier and Smith (1987) removed nest site shortage as a factor limiting nesting density. They carried out removal experiments and found that vacant territories were usually occupied by non-breeding birds and concluded that in the Bufflehead territorial behaviour limited nesting density and prevented some birds from breeding.

Figure 3.3 Two pairs of territorial Harlequin Ducks on the Laxa river, Iceland in May.

In ducks, territorial behaviour is most evident in riverine species, and this leads to regular spacing. Ball *et al.* (1978) studied the territorial behaviour of the African Black Duck in detail and found that about 90% of the territory was effectively defended against neighbouring pairs. Territorial spacing resulted in limitation of the breeding population; there was a high proportion of the population that did not breed. Black Ducks bred at different densities on different rivers; this reflected the abundance of food, pressure from conspecifics, or an interaction of the two.

Goosanders also spaced themselves along stretches of river in Canada (Wood 1986b). There was a close relationship between the size of territories and the availability of juvenile salmon (including those released from hatcheries along the river), despite the fact that the territories were established before any salmon were available. Wood suggested that Goosanders assess the quantity of food in streams the previous year and established their territories in the following spring based on their previous experience.

3.6 The egg and clutch

Waterfowl young leave the nest very soon after hatching and collect food for themselves. They hatch with substantial food reserves in the form of a

Table 3.2 The egg weight, clutch size and clutch weight of a range of waterfowl in relation to the body weight of the female.

Species	Female body weight (g)	Egg weight (g)	Clutch size	% Body weight	
				Egg	Clutch
Mute Swan	9,000	340	6.0	3.8	23
Black Swan	5,100	260	5.5	5.1	28
Snow Goose	2,500	122	3.9	4.9	19
Barnacle Goose	1,700	102	4.3	6.0	26
Upland Goose	2,800	120	6.0	4.3	26
Common Shelduck	950	80	9.0	8.4	76
Wood Duck	700	45	13.0	6.4	84
White-winged Wood Duck	1,900	89	10.0	4.7	47
Mallard	1,100	54	9.5	4.9	47
Blue-winged Teal	380	28	10.0	7.4	74
Canvasback	1,150	70	9.5	6.0	58
Tufted Duck	700	53	10.5	7.6	79
Common Eider	2,000	110	4.3	5.5	24
Velvet Scoter	1,200	80	9.2	6.7	61
Goldeneye	750	60	8.5	8.0	68
Goosander	1,200	85	9.4	7.1	67
Ruddy Duck	600	76	8.1	12.6	103
Maccoa Duck	750	88	5.0	11.7	59

yolk sac, which enables them to survive without feeding for up to seven days. In general, because of the needs of the precocial young the eggs of waterfowl are large relative to their body size, but within the group there is a great deal of variation (Lack 1968).

Table 3.2 shows the egg and clutch sizes for a number of waterfowl belonging to different tribes in relation to the body weight of the female. In general in birds, the size of the egg as a proportion of the body weight of the female decreases with body size, i.e. larger birds lay relatively smaller eggs. Swans and geese lay large eggs relative to their size; their young are large and mobile and are able to travel some distance without feeding. Note also that in this group, as well as in the Eider Duck, which use body reserves to a large extent for clutch formation and for incubation energy, the weight of the clutch as a proportion of body weight is much smaller than for other species.

The stifftails are exceptional in that their eggs are proportionately extremely large; the Ruddy Duck female lays the equivalent of her own weight in a single clutch. The young hatch with good reserves of fat which helps to insulate them in their almost wholly aquatic life and the large eggs are also resistant to chilling (Lack 1967).

Lack (1967) found that as a general rule in waterfowl, the larger the egg was in relation to body size of the female, the smaller the clutch, i.e. a species evolved to lay a large number of relatively small eggs or few large ones. Lack suggested that the size of the clutch was governed by the amount of food available to the laying female and the size of the egg. For geese, which carry food with them in the form of reserves, it is the size of those reserves that are important (Ryder 1970); this is only a slight extension of Lack's hypothesis. Johnsgard (1973), on the other hand, argued that since the eggs of swans and geese represent such a small proportion of their body weight, the need to produce large eggs was unlikely to set a limit on clutch size.

There are several alternative hypotheses to that of Lack, including the suggestion that clutch size is limited by the ability of the female to incubate a clutch successfully or to rear larger numbers of young. Studies involving the manipulation of clutches have been carried out in Canada Geese (Lessells 1986) and in Blue-winged Teal (Rohwer 1985). In neither of these studies was there any evidence that either the ability to incubate or to rear young limited the clutch size; the survival of eggs and young were not significantly reduced when clutches were artificially increased. The conclusion of both studies was that the size of the clutch was determined by the time that the first egg was laid.

Another suggestion, proposed by Johnsgard (1973) and others is that the chances of predation during laying, which increases with clutch size because large clutches are left unattended for longer, sets an upper limit on clutch size. Arnold *et al.* (1987) proposed that this would have some effect and also came up with a new suggestion, the egg viability hypothesis. They argue that since the chances of successful incubation decrease with the length of time that an egg is left un-incubated (for which there is some evidence), at some stage the decreased viability will counterbalance the benefit of laying additional eggs. There is little evidence, and some of it is conflicting, to support the predation hypothesis, and the egg viability suggestion has no empirical evidence to support it.

The amount of nutrients allocated to the clutch is also modified by the needs of the female herself for maintenance during incubation and there is clear evidence from Snow Geese that females with larger reserves lay more eggs, and that, whatever their original reserves, birds tend to retain a similar store of reserves for incubation (Ankney and MacInnes 1978, see Figure 6.1). Hamann *et al.* (1986) found that follicular atresia (the cessation of the growth of a follicle and the absorption of its nutrients) was a method by which individual females modify their clutch size after egg formation had started. Thus several follicles begin to develop and in conditions where the female obtains sufficient reserves, all develop into eggs. In unfavourable situations, however, one or more of the ova become atretic, allowing the female to adjust her clutch size to those of her reserves and leave sufficient for incubation.

Within the geese, the proportionate egg weight (the weight of the egg as a proportion of that of the female) increases with latitude (Owen 1980a, p. 114) and large young have been shown to survive better in cold conditions in Snow Geese (Cole 1979). There is a great deal of variation within a species; in Snow Geese the smallest egg was only 59% of the weight of the largest (Ankney and Bissett 1976). In the same study it was found that the total weight of some clutches of 4 eggs was greater than some with 5, and some 5 egg clutches weighed more than some with 6 eggs. The explanation was that in some (mild) seasons when large eggs did not confer an advantage, more young resulted from more (smaller) eggs, whereas the opposite was true in late (colder) seasons when survival of small young was low. The size of the egg, therefore, is crucial to the survival of young waterfowl and in general the evidence indicates that the clutch size is, as Lack originally proposed, determined by the availability of nutrients to the laying female, modified by egg size.

3.7 Incubation

Waterfowl nests contain a variable amount of down, the main function of which is to insulate the eggs during the female's absences during incubation. In some species it also serves to camouflage the nest during absences. Plucking down also exposes an area of skin—the brood patch—which is in contact with the eggs during incubation. The amount of down varies with the species; whistling ducks, some swans and stifftails pluck little or no down. The habit is most developed in geese and Eider Ducks, whose down has long been used commercially.

Thompson and Raveling (1988) tested the hypothesis that nest characteristics and the amount of down were related to exposure and incubation behaviour in three species of Arctic breeding geese in Alaska. They measured the rate of cooling of eggs in different positions in the nests of the three species and their results are shown in Figure 3.4. The Black Brant uses scant nesting material but has a profuse amount of down and during incubation is on the nest for 89.9% of the time; the Cackling Goose has a considerable amount of down and some vegetation and is 93.6% attentive;

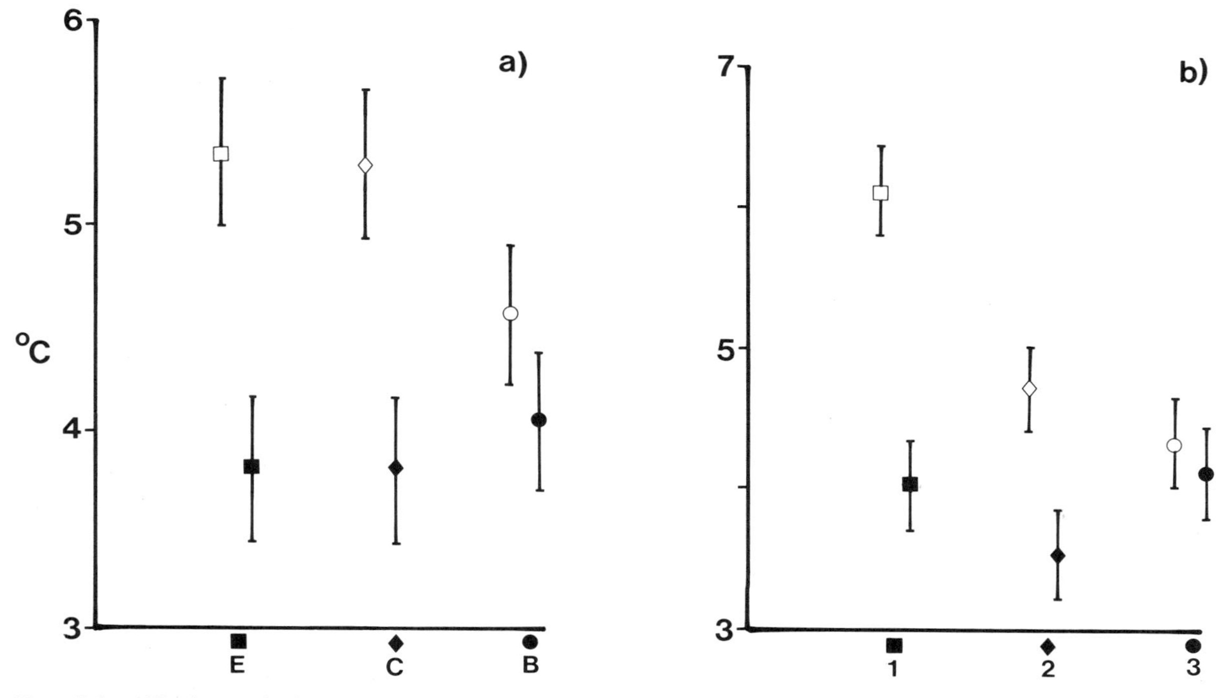

Figure 3.4 a) The decrease in the temperature of eggs of Emperor (E), Cackling Geese (C) and Black Brant (B) when left in nests in windy (open symbols) and calm (closed symbols) conditions. b) The combined average decrease in temperature of eggs of the three species in relation to the position of the egg in the nest 1—windward edge, 2—leeward edge and 3—centre. From Thompson and Raveling (1988).

the Emperor Goose is the most attentive (99.5%), and has a nest made up of mostly vegetation and little down. The results support the idea that the main function of the down is the insulation of the eggs during nest absences.

Incubation begins usually when the penultimate or final egg is laid, and lasts between 22 (Teal, Ross' Goose) and 36 (Mute and Black Swan) days. The incubation period depends largely on the size of the egg, though there are clear differences between taxonomic groups and between species breeding at different latitudes. The relationship is shown in Figure 3.5, using only a selection of the world's waterfowl. In general, Arctic species have shorter periods than expected on the basis of egg size and tropical species longer. After allowing for size, geese and swan embryos develop more rapidly than those of ducks. At the extreme the Ross' Goose (egg

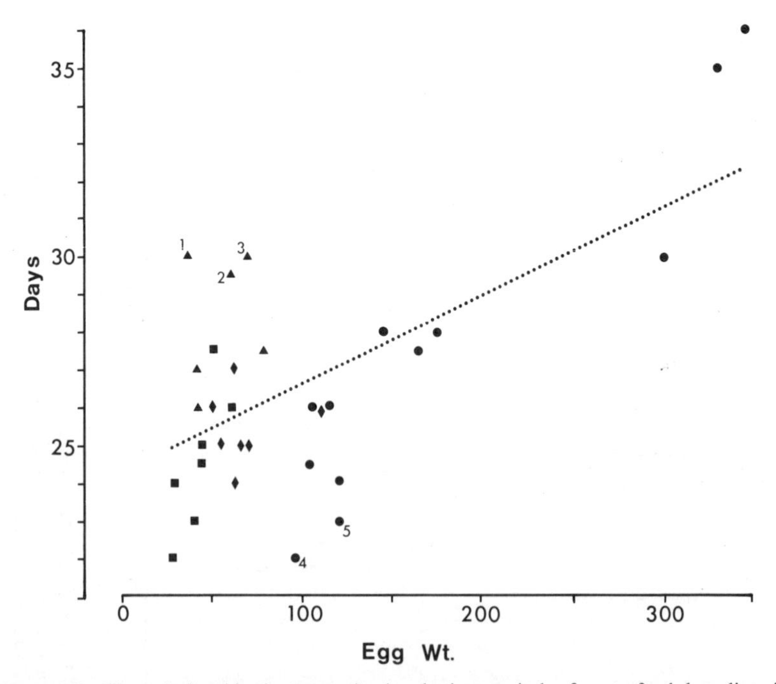

Figure 3.5 The relationship between the incubation period of waterfowl breeding in temperate and arctic regions of Europe (data from Owen 1977) and North America (Bellrose 1976) and the size of their eggs (weight in g). Dots—Anserini, squares—Anatini, diamonds—Aythyini and triangles—Mergini. The species showing most deviation are three hole-nesting *Bucephala* species (1–3), which have longer than expected incubation periods and two high arctic geese, Ross' (4) and the Lesser Snow Goose (5) with shorter periods.

weight 98 g) has an incubation period of twenty-two days, as does the Teal (egg weight 27 g). Among the geese, the white (Snow and Ross') geese, which have evolved in areas with very short seasons, are exceptional in having shorter than expected incubation periods, whereas in the sub-tropical Nene, the incubation period is exceptionally long (Owen 1980a, p. 115).

With the exception of the first four groups in Table 3.1, the female exclusively incubates. In swans, geese and sheldgeese the male guards the nest and territory, whereas in most ducks the female is left by her mate early in incubation and there is no active nest defence (Kear 1970). If the nest is threatened however, the female of some species will perform a distraction display to lead predators away from the nest.

The incubation constancy (the proportion of the time spent incubating) of waterfowl is generally high despite the fact that the incubating bird is not fed by its mate. Incubation behaviour has been studied in detail in a few species, including the Whistling Swan, one of the species exhibiting shared incubation (Hawkins 1986). The male sat on the nest during almost the whole of the period when the female was absent (20–40% in different females) thus protecting the eggs from predators as well as reducing the rate of heat loss (males do not have a brood patch).

Hawkins used dummy eggs to check the rate of egg cooling and found that eggs cooled on average 2.4° during female absences when the male sat. During periods when neither sex was incubating however, the eggs cooled 2.5 times more quickly than when covered by the male. The male does, therefore, have an important function in warming the eggs as well as protecting them from predators. Shared incubation allows the female to feed more during incubation and thus allows her to put more nutrients to the eggs than if she was the sole incubator. This may be crucial to the breeding success of swans, especially those which migrate large distances, carrying their reserves with them, to the breeding grounds. Further studies, for example on the relationship between incubation constancy and clutch size and body condition of females, could illuminate aspects of the energetics of breeding and success in swans.

The Emperor Goose is the most loyal to the eggs during incubation of all the geese for which information exists (Thompson and Raveling 1987). Despite the fact that the female alone incubates, she sits on average 99.5% of the time, taking, on average, a recess every two days. This extreme constancy, Thompson and Raveling argue, is related to the vulnerability of the nest to predation by birds during the female's absence, whereas an occupied nest can be defended against both avian and mammalian predators. The Emperor arrives on the breeding grounds with very large energy reserves and so can lay a clutch of 5 or 6 eggs and still retain enough energy for incubation with minimal feeding. The species travels rather short distances between wintering and breeding areas and so has a considerable advantage over more southerly wintering species.

Blue-winged Teal studied in northern USA were absent from the nest for an average of 4.8 hours per day—20% of the time (Miller 1976). The female fed for 60% of the time during recesses, indicating that the requirement for feeding is high during incubation. Ducks and geese, therefore, have a different strategy for the use of reserves for laying and incubation. Geese rely largely on reserves during the entire period, whilst ducks devote more to eggs and rely on feeding during incubation. The difference arises because of the different effect of occupation on the predation rate. Goose nests are usually in the open and vulnerable to predation during absences. Occupation deters aerial predators, and in the larger species, mammals as well. Duck nests are usually under cover of vegetation and the presence of the female makes little difference to mammalian predators. Thus occupation of the nest by geese makes a substantial difference to its vulnerability whereas that in ducks does not. Swans appear to be intermediate; the predation threat is counteracted by shared incubation.

An exception to the rule in ducks is the Eider Duck. The Eider is similar to geese in its nest sites—it breeds on islands inaccessible to land predators and nests in the open; the presence of the female deters nearly all avian predators. The Eider also behaves energetically like a goose, laying relatively few eggs, being almost 100% constant to the nest during

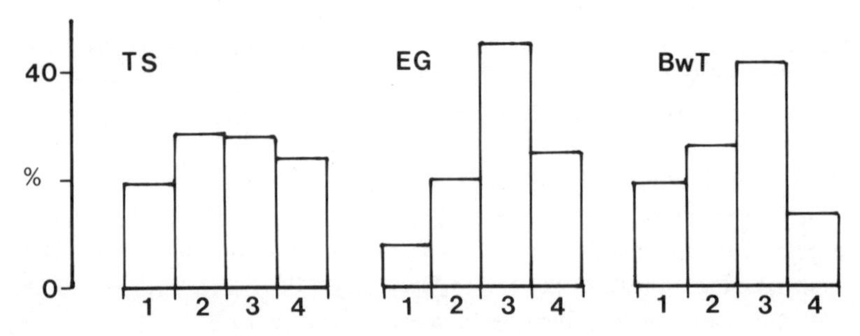

Figure 3.6 The proportion of the nest recesses of incubating waterfowl which began during the following periods of the day. 1—0001–0600; 2—0601–1200; 3—1201–1800; 4—1801–2400. Whistling or Tundra Swan from Miller (1976), Emperor Goose from Thompson and Raveling (1987) and Blue-winged Teal from Hawkins (1986).

incubation, when relying entirely on reserves for maintenance (Milne 1976).

Female waterfowl vary their incubation behaviour according to the environmental conditions; the pattern of absences during different parts of the day for a duck, goose and swan, all nesting in the wild, is shown in Figure 3.6. All tend to leave the nest during the warmest part of the day, when egg cooling would be slower; note that the difference is least in the Whistling Swan, where the cooling of the eggs is moderated by the incubating male during the female's absence. There was a positive correlation between the length of absences and ambient temperature in the Blue-winged Teal (Miller 1976). In southern Canada, where high summer temperatures might cause the eggs to overheat, female Mallard increased their attentiveness during periods when temperatures exceeded 32° C (Caldwell and Cornwell 1975). In the same study, which was carried out in captivity with easily available food, females sat for 94.6% of the time, reflecting the lessened pressure for food finding.

3.8 Renesting

Waterfowl normally nest once in every calender year and if nesting is successful, the birds complete the breeding cycle and usually moult. If, however, the first nest is lost during laying or early incubation, females of some species can lay a new clutch. Renesting is unknown in swans and rare in geese in the wild. This is because of the difficulty in gaining sufficient nutrients to lay a new clutch and because of the limited length of the breeding season rather than a physiological inability to lay; most species re-lay freely in captivity.

Most ducks can renest if the first nest is lost during laying or early in incubation; the relationship between the stage at which the nest was destroyed and the time to re-laying (renesting interval) is shown in Figure 3.7. The minimum interval is usually 4–5 days (minimum 3) and it increases by 0.62 days per day of incubation. Beyond a certain incubation stage relaying is impossible. Exceptionally, a female will lay a third clutch if the first two are lost.

Second clutches are usually smaller than the original laying; the difference in the size of the first and second clutch with supplementary food is, on average, about half an egg. The capacity of females to renest is also related to the food supply; experiments in captivity showed that females without supplementary food laid smaller clutches at longer intervals than

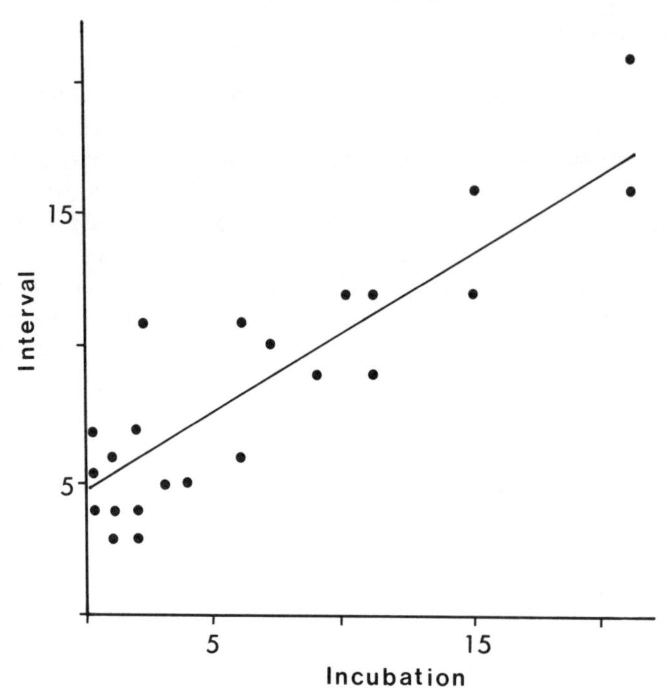

Figure 3.7 The relationship between the stage of incubation when the first nest was destroyed (in days) and the interval to the laying of the second clutch (renesting interval in days), for a number of duck species nesting at Delta, Manitoba. From Sowls (1955).

supplemented females under otherwise the same conditions. (Swanson *et al.* 1986). The amount of yolk and of total lipid in second clutches were significantly less than in first layings of the same individual Barnacle Geese in semi-captivity (Owen and West 1988). Thus the nutritional constraints affect egg quality as well as clutch size, renesting capability and interval.

3.9 Hatching and rearing

In the last three or four days of incubation the young start calling inside the eggs and the female increases her attentiveness to the nest. Males of those species that defend during incubation also usually move closer to the nest. Communication between the young helps to synchronise hatching—very important in precocial birds to ensure that all the chicks can leave the nest together. Weak chicks have to be abandoned in the nest as the need of the remainder of the brood to find food becomes over-riding.

Young waterfowl exhibit a following reaction and are led by the female or both parents to the feeding grounds. Species nesting in elevated positions have first to jump onto the ground (in hole nesters, having first climbed out of the hole). An experiment with a visual cliff demonstrated the differences in the adaptations of ducklings; young of tree nesters showed no preference for level ground over an apparent drop, whereas ground nesters stayed on the shallow side (Kear 1967). Tree holes used by diving ducks are usually between 1 and 5 m high, and the young of Barnacle Geese may have a drop of as much as 60–70 m from their cliff nests. The young are, however, light and fluffy and usually come to no harm from the fall.

All waterfowl are precocial, but there is a varying degree of parental care; a comprehensive review is given by Kear (1970). All young can feed for themselves, but a few species assist by presenting food (Magpie Goose and Musk Duck) and swans and some geese are known to stir up the bottom, bringing food up to the surface by 'foot trampling'.

Some waterfowl, especially swans, carry small young on their backs to brood them and protect them from predators, but the extent of most parental care is vigilance and some degree of defence against predators.

The most extreme degree of care is shown by the Anserini (see page 76). In Mute Swans the sexes moult asynchronously, which means that at least one parent can fly at all times (Birkhead and Perrins 1986), and male Barnacle Geese retain their flying abilities until the goslings are large enough not to be vulnerable to aerial predators (Owen and Ogilvie 1979). South American sheldgeese behave in a similar way to their ecological counterparts in the north, emphasising that parental behaviour is an adaptation to the ecological conditions—short breeding seasons, vegetarian diet and delayed maturity.

In some species young from different broods mix and form a creche when they are quite young; the creches are accompanied by a few adults. Creching is probably most well developed in the Shelducks, but it is not clear whether it has developed as a strategy to increase survival or whether it arises accidentally when many broods interact on the feeding grounds (see page 73).

In most ducks, only the female accompanies the ducklings until they can fly, usually at 5–10 weeks. In a few species, including the steamer ducks of South America, both parents are in attendance. In the Magellanic Flightless Steamer Duck on the mainland, the male is always with the brood, whereas in the same species in the Falkland Islands, male attendance is 91%. This is reduced further to 72% in the Flying Steamer Duck. The degree of parental care, as with other characteristics is an adaptation to the harshness of the environment in which they breed (Weller 1976).

In the Oxyurini, the young become independent before they are fledged. In the North American Ruddy Duck, the distance between ducklings and their nearest and furthest neighbours increased with age and most broods were abandoned by the female after the fourth week (Siegfreid 1977). In the exclusively parasitic Black-headed Duck, the duckling is independent of parental protection from the first day (see page 75).

3.10 Growth and development of young

The growth rate of young waterfowl varies mainly in relation to the quality and availability of food. In geese and swans this is related to the breeding latitude. In temperate Europe, young Mute Swans take 120–150 days to reach the flying stage (Birkhead and Perrins 1986), whereas similar sized Trumpeter Swans in Alaska fledge at 90–105 days (Bellrose 1978).

Variations with latitude are at least partly due to the fact that the daylight available for feeding in summer increases with latitude. The fledging period of young geese reared under identical conditions of continuous daylight in captivity depended almost entirely on body size (Figure 3.8). The correlation is very close, but the deviations do conform to the general hypothesis of increasing growth rate with latitude. For example, the Nene (Y) breeds at the lowest latitude and also has the slowest growth rate. The low-latitude Swan Goose (A) and the Canada Goose races (V, W, X), which generally breed at lower latitudes than expected on the basis of their size also grow more slowly than expected. Within species the same applies; the Western Bean Goose (B) breeds at 66° N, whereas the Russian Bean (C, 72° N) has an earlier fledging time. The same is true of the Greenland White-front (F, 68° N) and its Russian counterpart (E, 73° N).

The growth curve of ducklings is similar in all species studied (all in captivity); generally fledging takes 5–10 weeks. Street (1978) varied the quality of the diet of Mallard ducklings by including different proportions of protein-rich animal food. There was a very close positive relationship between protein content of the diet and growth (correlation between protein percentage and body weight change in the first four days, $r = 0.94$). The amount of insects available for wild ducklings clearly affects their growth rate and chances of survival. There is also some evidence that the

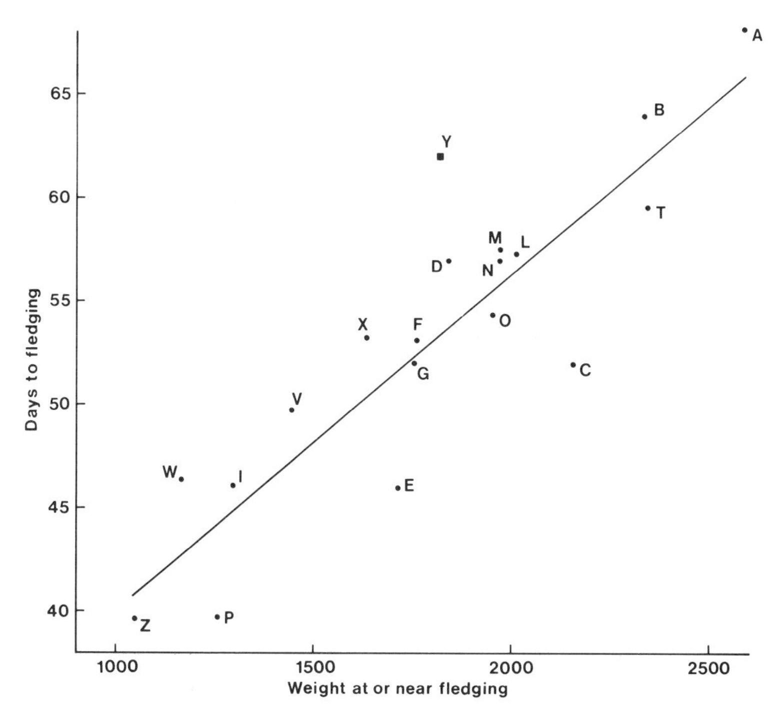

Figure 3.8 The 'fledging period' of geese reared in captivity under conditions of continuous daylength in relation to the body weight (g) at or near fledging. Goslings were regarded as having fledged when the tips of the primary feathers reached the base of the tail feathers. Letters refer to different species; the square (Y) refers to the Nene. Including all points correlation coefficient $r = 0.914$, $P < 0.001$. From Owen (1980a). For explanation of letters see Appendix.

quality of the rearing diet affects final size. Smart (1965) found that the wing lengths of late-hatched Redhead ducklings were shorter at fledging than those of early broods. Owen and Montgomery (1978) found significant differences is the wing lengths of juvenile and adult Mallard in different seasons and suggested that the difference related to seasonal differences in the earliness of nesting. This is consistent with the idea that early breeders benefit by their young hatching at the time when the quantity and quality of food is highest.

3.11 The effect of age on reproduction

In most animals reproductive performance increases with age and experience and there are indications from some species that fecundity declines in old age (see Clutton-Brock 1988). This effect influences breeding at all stages; the non-breeding rate decreases with age, clutch size and nesting success increases. In the Mute Swan it is the age of the male that is most important—the correlation between the number of young fledged and male age was positive and significant. Older males bred earlier and their mates laid larger clutches, and clutch size was the most important factor influencing the number of cygnets fledged (Birkhead *et al.* 1983).

Laying date is closely related to the ability of the male to put on weight in spring, and presumably to the quality of the territory; young Mute Swans spend some years without a territory and having established one do not usually breed for the first year (Birkhead and Perrins 1986). It is difficult to separate the effect of age from that of breeding experience, but in Mute Swans the difference in the productivity of experienced pairs was greater than would have been expected on the basis of age difference alone (Birkhead *et al.* 1983), so both age and experience are probably important, at least in swans. In Bewick's Swans, the size and fighting ability of the male also appear to have an effect on a pair's breeding success (Scott 1988).

There is ample evidence that age affects all aspects of breeding in geese, and the results of two long-term studies which have examined the success of individually marked birds are shown in Figure 3.8. The clutch size of Snow Geese increases with age up to five years and remains stable thereafter, at least up to the age of nine. Recruitment for several cohorts however, declines in old age; reproductive performance peaked at eight years of age (Ratcliffe *et al.* 1988). The measure of breeding success used in the Barnacle Goose study was bringing young to the wintering grounds, and this shows an increase in productivity in the first years of breeding life and also a decline after the peak reproductive life of 7–12 years.

The body weights of female Eider Ducks (which have a similar breeding strategy to geese) at the start of laying is greater in older birds, which also breed earlier. It is suggested that the age effect in this species is related to the male's ability to defend its mate during the pre-laying and laying period (Baillie and Milne 1982). In geese, the decline in production in old age may also act through a reduction in fighting ability which results in a smaller proportion of older pairs establishing territories and raising broods in competition with pairs, and particularly males, in the prime of life. Data from the Barnacle Goose study (Figure 3.8) add support to this. The production of a group of older geese which lived at a time when the population was small and increasing, and when there was much less competition for resources, was considerably higher than the group that lived in years of high density. There were confounding, area effects,

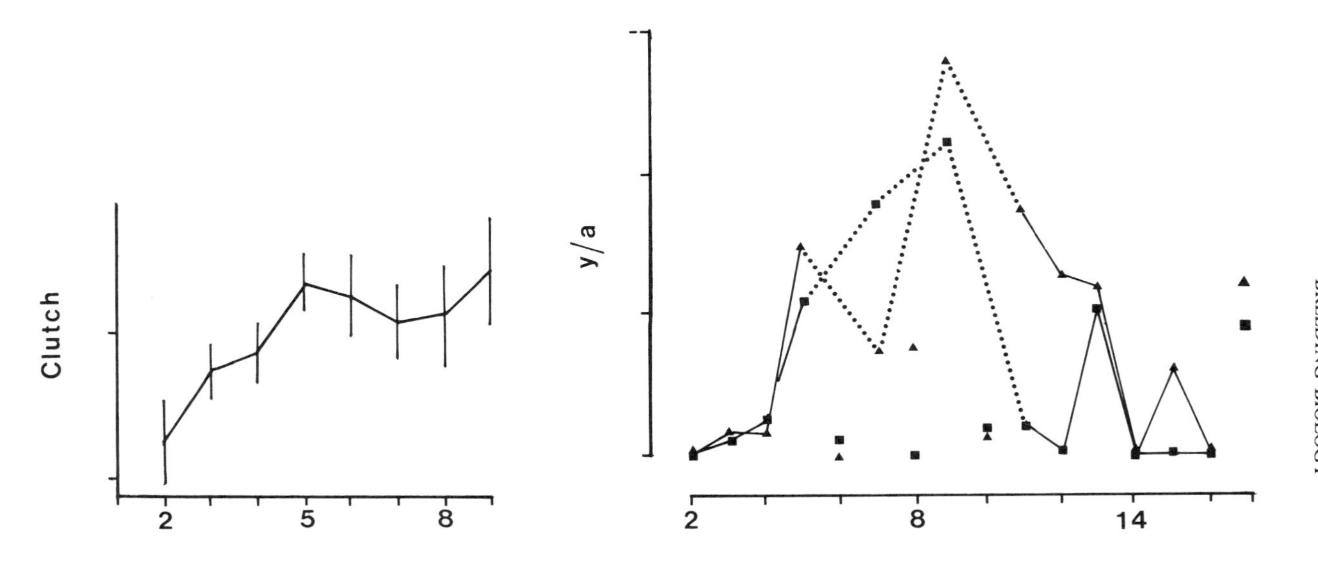

Figure 3.9 The clutch size of Lesser Snow Geese breeding at La Perouse Bay, Canada in relation to the age of the female (left—from Rockwell *et al.* 1983). The reproductive rate (number of young produced per adult of a cohort of Barnacle Geese through their lives (right)). Squares—females and triangles—males. Very poor breeding years (1977, 1979 and 1981) are omitted and the points for the good years joined with dotted lines. Sample sizes at the start n = 41 females and 37 males. The points at extreme right are from a different group of geese which are at least 17 years old. From Owen and Black (1989b).

Table 3.3 The relationship between the age of the female and a number of measures of fecundity in the Lesser Scaup. The results are mean, standard errors (and sample size) values for females of different age over a period of four years (not the same females in each year), so some of the variability relates to between-year effects. From Afton (1984).

Parameter	Female age (years)			
	1	2	3	4
Arrival date (Julian)	129.1 ± 2.3 (37)	127.0 ± 2.9 (12)	132.2 ± 3.2 (9)	122.8 ± 2.4 (12)
Non-breeding	29.3% (58)	9.6% (52)	0.0% (23)	0.0% (23)*
Arrival–laying (days)	36.6 ± 3.3 (9)	39.3 ± 2.2 (6)	29.9 ± 2.7 (9)	37.4 ± 2.3 (9)
Clutch size	9.0 ± 0.1 (26)	10.0 ± 0.2 (21)	10.9 ± 0.3 (16)	12.1 ± 0.2* (14)
Nest success	26.3% (38)	22.2% (45)	45.5% (22)	41.7% (24)†
Renesting	8.7% (23)	17.9% (28)	27.3% (11)	18.2% (11)
Brood survival	0.69 ± 0.10 (10)	0.67 ± 0.11 (10)	0.58 ± 0.12 (9)	0.75 ± 0.07 (10)

*Significant, $P < 0.001$; †$P = 0.0554$.

however, geese that were younger than the cohort illustrated in Figure 3.9 bred better because they nested in an area providing better food during incubation and rearing (Owen and Black 1989b).

Similar relationships are also found in ducks, as summarised in Table 3.3 for the Lesser Scaup. Arrival date and the pre-laying period are not related to age, but clutch size increases progressively at least to 4 years of age. Without considering non-breeding, and combining all other effects, we can calculate that the number of young produced by an average 1-year-old female would be 1.64 young, compared with 3.76 ducklings by 4-year-old and older females. Including non-breeding (absent after the second year), the yearling production is reduced to 1.16. The propensity for renesting is also higher after the first year. Detailed studies of Mallard in captivity indicated that renesting effort increased in the first three years (Swanson *et al.* 1986). In Goldeneyes, as well as age, experience of the nest site and its surroundings is advantageous; females returning to the same site had higher success than those which moved to another nest box (Dow and Fredga 1983).

Since age of the parents affects the chances of survival of their young, at least in some species (see Chapter 6), the contribution of birds to the productivity of the population in waterfowl is heavily biased in favour of the older age groups.

3.12 Lifetime reproductive success

Because of the difficulty of following animals over the whole of their lives, studies which measure lifetime reproductive success (LRS) are rare, subject to several sources of bias, and samples are usually small. There are only a few waterfowl species which have been studied closely enough to provide estimates of LRS, notably the Bewick's Swan (Scott 1988) and Barnacle Goose (Owen and Black 1989b). In swans, the presence of the mate and its lifespan were very important; mate loss caused a reduction in success and affected lifetime reproductive success.

In Barnacle Geese the main determinant of LRS was longevity, and a difference in the lifespan of males (median 10 years) and females (8 years), was reflected in a comparable difference in LRS (mean lifetime production in a cohort born in 1972 was 2.6 ± 0.44 young for males and 1.5 ± 0.29 for females). An important factor affecting LRS of an individual was the weather conditions during its lifetime. Using average breeding performance for the whole population, we calculated that individuals of median lifespan which lived through climatologically the 'best' period could produce more than 40% more young in their lives than those which lived during the 'worst' period.

In the 1972 cohort, 35% of males and 49% of females did not breed successfully at all during their lifetimes and only 15% of birds contributed half the next generation's recruits. In that Barnacle Goose population, average breeding success fell as numbers increased and competition for resources intensified. Since brood size did not decline, a smaller proportion of successful breeding efforts were contributing to the population's productivity. It is likely also that the proportion of birds failing to breed in their lifetimes and the proportional contribution of the most productive geese increase with population size.

It has been proposed that in order to maximise lifetime reproductive success, an individual should adjust its reproductive effort in any one season according to its chances of breeding successfully (Williams 1966). This assumes that there is a trade-off between current reproductive success and the chances of survival and future breeding. There is very little information on these aspects but Lessells (1986) was able to study the effect of brood size on adults by manipulating brood size in Canada Geese. She found that the weight of females at the start of moult and the timing of their moult were negatively correlated with brood size, but found no evidence that this affected survival. Females with larger broods laid later in the following year, though their brood size was unaffected. Schindler and

Lamprecht (1987) presented evidence from Bar-headed Geese that brood size affected the time budgets of parents during early rearing, and there is similar evidence from Barnacle Geese (Black unpublished). In that species, Black and Owen (1989b) could find no evidence that successful breeding in one year affected the chances of survival to, or of successful breeding in the next, for either sex. In fact there were indications that pairs that cared for their offspring longer were more likely to return with young the following year. This relationship is however, not likely to be causal; more probably it is the result of a correlation between the attributes of successful birds. The evidence for a trade-off between current and future reproduction is conflicting and more long-term studies are required to verify the hypothesis.

3.13 Flightless moult

In common with other water birds, waterfowl moult all their primary and secondary flight feathers simultaneously and become flightless for a period. The moult follows closely after breeding and occurs on or close to the breeding grounds. Immature geese and swans, although they do not breed, travel to the breeding area and moult there. Not all species are well studied, and there are exceptions to the pattern of an annual flightless period. The Magpie Goose, for example, moults sequentially and remains capable of flight all year round. Sheldgeese, according to recent evidence, may also be exceptions, at least in some circumstances. Summers (1982) found that Ruddy-headed Geese in Argentina had old and new flight feathers at the same time. They tended to moult primary and secondary feathers at different times and some birds showed evidence of sequential moult, involving the shedding of one or a few feathers at a time. The Greater Magellan Goose normally has a flightless moult, but in some years, some individuals 'skip' a moult and retain their primaries for two cycles (Summers 1983). The only evidence for such a phenomenon in northern species involves a single Whooper Swan which had its primary feathers dyed in one winter and returned in the following year with the same dyed feathers (Campbell and Ogilvie 1982).

The duration of the moult varies with the species and the length of the primaries. In Mute Swans it takes 66–67 days for the flight feathers to grow to full length, though the birds are able to fly before then (Matthiasson 1973). The Barnacle Goose is flightless for an average of 25 days (Owen and Ogilvie 1979) and the Mallard for 32–34 days (Owen and King 1981). The birds move to moult in secluded locations and stay on or near water to avoid predators. The male and female of some species moult at different times to provide protection to the family, as already mentioned.

It has traditionally been assumed that the moult is a stressful process, and that the decline in body weight during moult which has been reported for many species indicates that the birds are stressed because they are unable to find sufficient food (see e.g. Hanson 1962). Evidence is accumulating, however, that this is not the case. Douthwaite (1976), studying Red-billed Teal moulting in Australia found that some body reserves were present at the end of the flightless period, despite the fact that the birds lost weight during moult. He suggested that weight loss was adaptive in that it reduces the flightless period. Evidence in support of this comes from Barnacle Geese (Owen and Ogilvie 1979) and for Snow Geese (Ankney 1979); the weight loss is adaptive rather than imposed on the birds through food shortage and energetic demands. Mallard moulting in enclosures at Slimbridge and provided with a superabundance of energy-rich food nevertheless lost weight during moult (Owen unpublished). Eider Ducks, however, which are maritime during moult put on weight (Milne 1976). Presumably the absence of predation risk at sea reduces the selective pressure for early regaining of the powers of flight.

It seems, therefore, that at least in most species, waterfowl lay down body reserves prior to moult and lose them, whether or not food is abundant. The post-moult period, in the autumn, is usually the period of greatest food abundance following the main growing and breeding period of prey, so the birds are able to regain reserves quickly before the onset of winter. The selective advantage of reducing the flightless period outweighs the rather minor benefit of maintaining body reserves through the moult.

CHAPTER FOUR

SOCIAL AND SEXUAL BEHAVIOUR

A number of classic studies of a broad range of species have revealed that the mating systems—the range of behavioural and physical attributes that contribute to the process of producing offspring—are closely linked with the environment in which they live (Vehrencamp and Bradbury 1984). Social behaviour is the manifestation of the interplay of these attributes between individuals within a population.

Social and sexual behaviour in waterfowl in relation to the evolution of the group and taxonomy has been the subject of detailed and extensive study since the work of Lorenz (1941) and Johnsgard (1965). The ecological significance of social groupings and behaviour and their influence on food-gathering and ultimately breeding success or survival have been fully realised more recently, although Boyd's (1953) work on geese was an early example. In this chapter we focus only on those aspects of social behaviour that have direct ecological implications. Spacing and territoriality during the nesting season is discussed in Chapter 3; the observations here refer mainly to the non-breeding season.

4.1 Mating systems

Waterfowl exhibit a variety of mating systems, the predominant **monogamy** (exclusive pairing of one male and one female for at least one breeding season and sometimes for life), the strategy adopted by 90% of all bird species (Lack 1968). When more than one female is attended to by one male the system is **polygamy.** There can also be a mixed strategy.

Several models have been developed concerning the evolution of these systems (see Oring 1982 for review). The major selecting factors are the type and quality of the habitat and the sex ratio; the mating system is governed by factors that affect the chances of finding mates and resources. Waterfowl are an excellent group of birds to make mating system comparisons to test the ecological consequences of different strategies.

4.1.1 *Monogamy*

The length of time monogamous pairs remain together varies considerably in waterfowl. On one extreme there are the northern swans and geese which form strong pair-bonds where the partners are in proximity 24 hours a day and throughout the year. The bonds can last for years; the longest records so far from the wild are 19 years for Bewick's Swans and over ten years for Barnacle Geese.

Ducks are normally seasonally monogamous; pair-bonds last only a few months. Typically, pairs form on the wintering grounds or on spring migration and partners remain together until males desert the female some time during incubation (Sowls 1955). Some ducks re-pair with each other in the same breeding season to produce replacement or second clutches (serial monogamy) but in only a minority of duck species (usually in Tadornini and the long-lived members of the Mergini) do birds re-pair with the same mates in successive seasons. In some species, such as the Barrow's Goldeneye, pair members find each other when they return to the same wintering sites (Savard 1985). Some highly territorial Southern Hemisphere species, such as riverine ducks, do remain together and have stable and long-lasting bonds (see page 81).

For males the main benefit to maintaining a pair-bond with a single partner is to ensure his paternity. In many ducks males actively repel neighbouring males particularly when the female becomes receptive prior

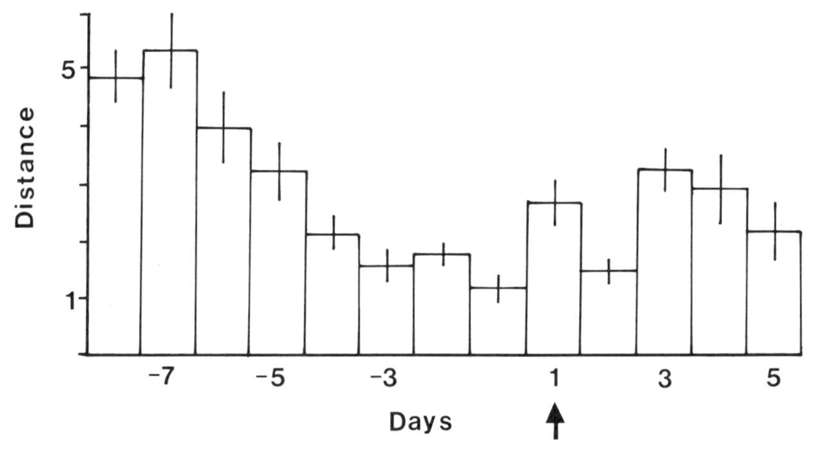

Figure 4.1 The mean distance (in bird lengths) between mates in eight pairs of Mallard before and after the start of laying (day 1, arrowed). From Goodburn (1984).

to nesting—mate guarding (Figure 4.1). Copulations between duck partners also occur before arriving on the breeding grounds presumably as part of the testing process in mate choice (McKinney 1986). Benefits for females after pairing with a mate relate to protection from disturbance to their feeding (usually by unattached males) ensuring adequate body reserves in preparation for egg laying and incubation. The advantage is clearly seen in the Eider Duck, and is an important advantage of the pair-bond for that species (Ashcroft 1976).

The advantage of the presence of the male to defend the female's feeding resources during the pre-breeding period has been clearly demonstrated in Arctic nesting geese which need to lay down substantial fat and nutrient reserves to enable success in breeding. On the spring fattening area in the Netherlands Teunissen *et al.* (1985) monitored how successful different male–female partnerships were in mate protection and in obtaining food. By recording detailed measurements on male behaviour and female feeding performance they found negative correlations between male aggressiveness and female walking rate through feeding patches and interruptions in feeding bouts by being displaced by neighbours. There were also positive correlations between male aggressiveness and female feeding rate (pecks/step) and female feeding time (Figure 4.2). The link between individual males' aggressiveness during flock feeding situations and the pairs' future breeding success has been established in wild and semi-captive Barnacle Geese (Black and Owen 1987, 1989b).

In most long-term monogamous species, the male takes a substantial role in rearing the young. One of the primary factors promoting long-term monogamy is thought to be the necessity of two parents successfully rearing offspring. This contention was tested in the large colony of Lesser Snow Geese at La Perouse Bay, Canada by removing males and females at various stages of the breeding process (Martin *et al.* 1985). These manipulations revealed that the partnership was essential in order to establish territories; neither the male or female could defend and maintain a nest site alone. Once the incubation stage was under way some females did succeed in hatching their eggs after their males were removed, despite varying degrees of harassment from neighbouring males. This experiment shows that although it is possible for female geese to rear offspring without the males' assistance, as do most duck species, at least in a colony situation being single is not the best option. Little is known about the post-fledging survival of young of single parent families or of their contribution to later breeding, both of which could be affected, for example if nutrition during rearing was limited.

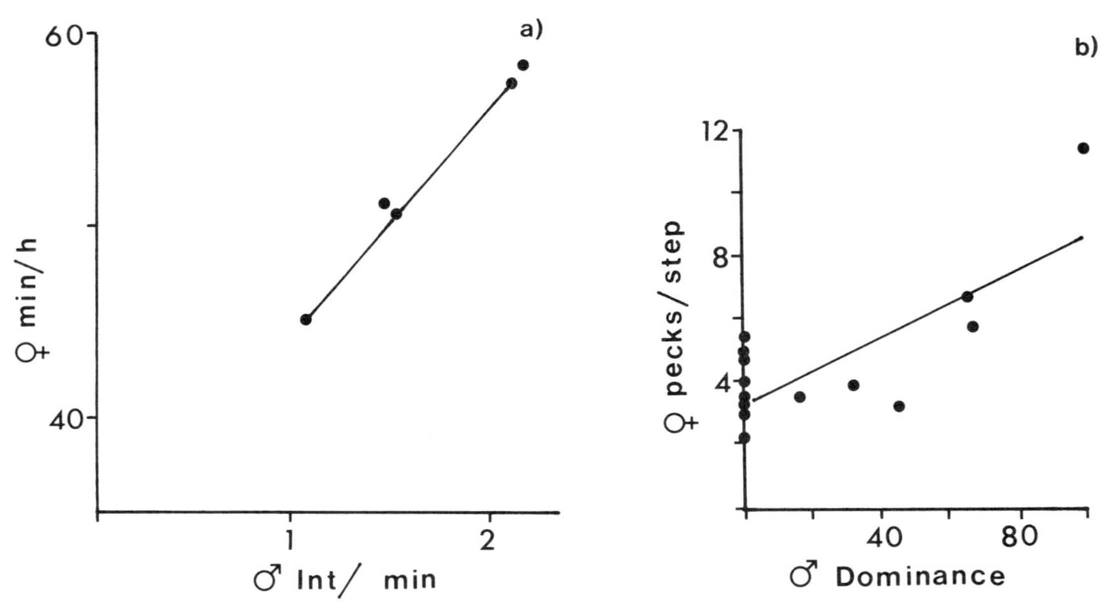

Figure 4.2 a) The foraging time (minutes per hour spent feeding) of female Brent Geese in relation to the aggressiveness of their mates, measured by the frequency of interactions with conspecifics. The points indicate means for a season for individual pairs (more than 2 hours of continuous observation). b) An index of the feeding rate of the female of a pair (the number of pecks taken per step) and the status of her mate, as assessed by his success in encounters with other geese in the foraging flock. From Teunissen *et al.* (1985).

The most revealing evidence that the monogamous system yields benefits in terms of natural selection can be seen from the results of the long-term studies on Bewick's Swans (Scott 1988) and Barnacle Geese (Owen *et al.* 1988). In Bewick's Swans, the change of a mate whilst both partners are still alive ('divorce') is extremely rare. No incidence was recorded among 500 pairs of Bewick's Swans which had bred successfully and only 20 cases among 1000 pairs which were never observed with broods. The lifetime production of individual swans was affected by mate changes; the mean number of cygnets raised in a year was significantly correlated with pair-bond duration.

The results for the very large samples of Barnacle Geese are given in Table 4.1. The difference in breeding success was highly significant in the first year after changing mates but there was no difference in the second year. Results from this study also suggest that the partnership between individuals has to mature before successful breeding attempts occur. Evidence for this comes from the finding that the pairs that re-form in the autumn or winter have a better chance of breeding successfully than those which form closer to the breeding season in spring. Many observers have noted that newly established pairs expend more time and energy in courtship and copulation than do old established mates and this could have a crucial effect on fattening rate and hence breeding potential in the wild. Several lines of evidence for Barnacle Geese have indicated that 'divorce' is

Table 4.1 The breeding success of 'divorced' birds with their old mate, and in the first and second year with their new mate, and the success rate of a larger sample of geese which includes those whose mates had died. Some birds remained unpaired for some time after divorce or mate loss. The figures in parentheses do not include unpaired birds. The records include all pair/years with the old mate and include both members of a divorced pair (a total of 35 separations, 65 birds surviving beyond year 1 after divorce and 49 beyond year 2). From Owen *et al.* (1988).

Breeding success	Old mate	New mate Year 1	New mate Year 2
Divorced birds			
Successful	32	5	8
Paired and failed	124	52	41
Unpaired (failed)	—	8	0
% Successful (of those paired)	20.5	7.7 (8.9)	16.3 (16.3)
All changes			
Successful	2392	30	87
Paired and failed	8822	307	319
Unpaired	—	20	4
% Successful (of those paired)	21.3	8.4 (8.9)	21.2 (21.4)

due to accidental separation and rapid re-pairing and Owen *et al.* concluded that most geese of all species probably strive for lifelong monogamy, which has advantages in terms of breeding success. The pair-bond is so strong that individual Barnacle Geese retain the same mate throughout life despite the fact that some geese do not breed successfully in as many as 16 seasons (Owen and Black 1989b).

4.1.2 *Polygamy*

To date only three waterfowl species are known to be predominantly polygamous, all of which are from tropical or subtropical areas of the Southern Hemisphere: The Magpie Goose (Frith and Davies 1961), the Maccoa Duck (Siegfried 1976) and the African Comb Duck (Siegfried 1979). The lengths of breeding seasons in tropical climates are often prolonged, enabling males to mate several times as females become fertile and produce several broods (see Chapter 3). Siegfried (1976) described the unusual behaviour of the Maccoa Duck which inhabits the wetlands of southern Africa. In this species males defend discrete territories and display to females that are prospecting for a nest. One male can have several females nesting in his territory. Many males occupied the fringes or inferior territories and were ignored by females.

In an experiment using birds in captivity Siegfried (1985) discovered that the bright plumage and cobalt-blue bill of the male Maccoa changed to a non-conspicuous colour much like the females' not only during the moult but also under stressful social situations. When a group of four males were kept in moderately sized pens with one female, only the most aggressive male had bright plumage. The subordinate males had dull plumage even though they were mature. When the dominant bird was removed another male became dominant and his plumage turned bright. In another pen four brightly coloured males and one female were introduced. Within four weeks their social rank was established and three of the males began losing their bright colour. In such a system dominant males probably acquire the majority of copulations whereas subordinates may gain access to females by adopting dull plumage while remaining on the fringe of male territories. Status signalling using plumage characteristics is rare in waterfowl. However, a link between age, social status and plumage characteristics is suspected in the White-headed Duck's variable plumage patterns (the amount of white on the head) and brilliant blue bill (Torres and Moreno 1986).

In a small flock of semi-captive Bar-headed Geese kept in Seewiesen, West Germany, failure of some males to return resulted in a 1.23 female to male sex ratio (Lamprecht and Burhow 1987). In this unequal situation over 10% of the males were followed by at least two and sometimes up to five females—a system known as harem polygamy. Secondary females were not protected by the male and were attacked by flock members more than primary females and spent less time feeding due to a heavier vigilance burden. Despite these disadvantages secondary females fared better than lone females. Three secondary females became the primary mate of their male immediately after the first female disappeared. Such a system has not, however, been recorded from wild geese.

Polygamy is very much the exception to the rule in waterfowl. The opportunity for access to more than one mate when females are fertilisable, together with the short duration of the breeding season in many species and aspects of behaviour such as territoriality, appear to be the main factors selecting against polygamy and promiscuous behaviour.

4.1.3 *Mixed reproductive strategies*

In certain circumstances female waterfowl may opt for a more flexible breeding strategy by laying her eggs in another female's nest (**nest parasitism** or **egg dumping**). In some ducks this tactic is a secondary reproductive strategy—a means of reducing the risk of total reproductive failure by females who do not have the opportunity to secure their own nests (Weller 1959). Failure to establish a nest may be due to low dominance status, lack of experience or to a limited number of nesting places and/or mates. For example, individually marked female Wood Ducks laid parasitically both before and after seasons when they laid only in their own nest site; the propensity for parasitism was independent of age. Some females laid a parasitic egg before establishing their own nest site and others dumped all their eggs in other nests (Clawson *et al.* 1979).

The tendency for parasitism in Lesser Snow Geese is linked to three factors. Young females are more likely to dump eggs, which may indicate the inability of young pairs to obtain territories. Females parasitised other nests more often in years of late snow melt when there were few nest sites available. Nest parasitism was prevalent in years when many females arrived on the breeding grounds in poor condition due to prairie drought during the spring fattening period. All three features indicated that dumping was a last resort strategy adopted when chances of normal breeding were slim (Lank *et al.* 1989).

In spite of the fact that most parasitic eggs fail to hatch due to poor laying synchrony with host eggs, a substantial number of young can result from parasitic eggs; 23% in the Wood Duck and 9% in the Lesser Snow Goose. A large number of studies have shown that hosts to parasitic eggs can suffer in terms of reducing their own clutch size, reduced egg hatchibility, higher nest desertion rates and reduced duckling survival (see e.g. Amat 1987b). However, some investigations on geese indicate that larger brood sizes are advantageous (see e.g. Black and Owen 1989b), offsetting the increased costs of rearing them.

The phenomenon of several females leaving their offspring with one or two other females (**brood amalgamation** or **creching**) is not uncommon (see page 57), but little is known about its function and evolution. In the Eider Duck, the risk of predation by gulls on ducklings decreases with creche size (Munro and Bedard 1977). In Barrow's Goldeneye, which rear their broods on the Laxa River in Iceland, roving females lose their broods to territory-holding females. Movements are linked to the timing of insect emergence which follows a gradient downstream. Territorial females become roving females when insect emergence declines and in turn lose their brood to other territory-holding females downstream (Einarsson 1987 and pers. comm.). Thus ducklings are left in areas with the richest food supply. In these studies it seems possible that females who give up their young to other females may benefit by increasing their offspring's chances of survival and by releasing themselves from further parental duties. These females may also enhance their own survival since they probably suffer less risk of predation when not accompanying broods. However, in most studies of ducks and geese to date it appears that the process of brood amalgamation is not adaptive but largely accidental when two broods cross paths or when families are in dispute.

Perhaps the most widespread mixed reproductive strategy in waterfowl is that of extra-pair **forced copulation.** It occurs regularly in most dabbling ducks, some diving ducks, stifftails and at least one goose species. It happens when a male tries to secure additional matings particularly after his primary mate has disappeared from the scene to incubate their clutch (McKinney *et al.* 1983). Sperm from these forced copulations are viable; up to 60% of Mallard ducklings came from broods that were multiply-fathered as assessed by electrophoretic techniques (Evarts and Williams 1987). Forced copulation is not costly for males and such behaviour would be expected since it maximises individual productivity (Trivers 1972).

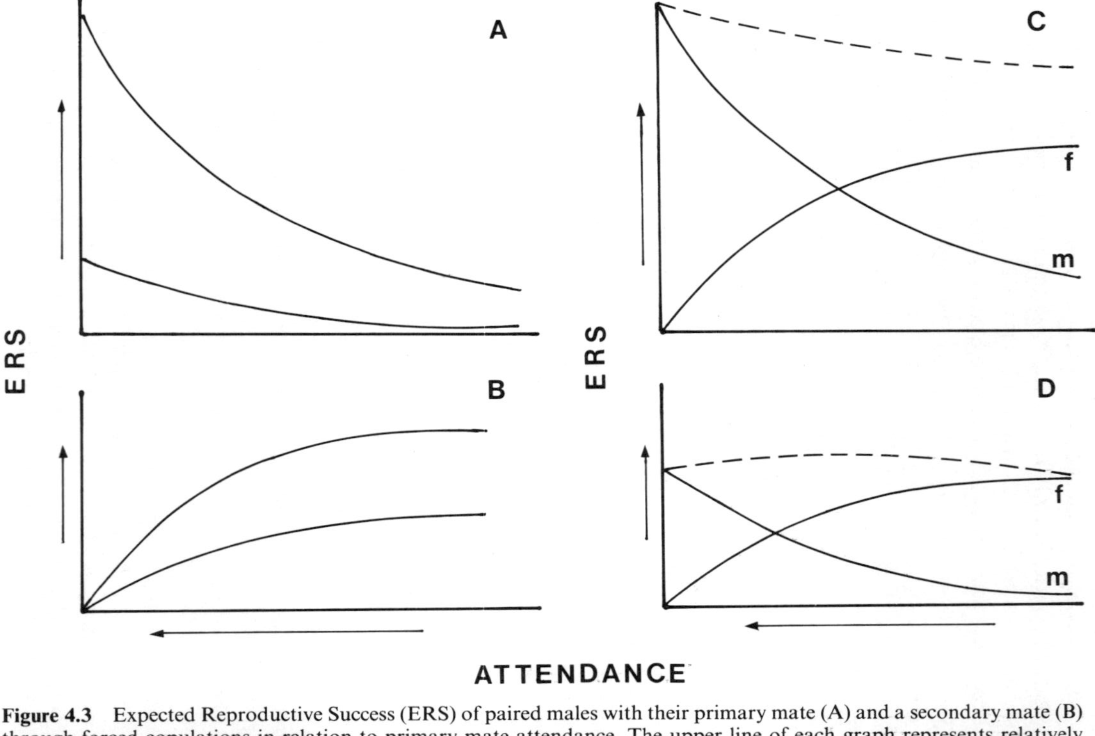

Figure 4.3 Expected Reproductive Success (ERS) of paired males with their primary mate (A) and a secondary mate (B) through forced copulations in relation to primary mate attendance. The upper line of each graph represents relatively stable, and the lower line relatively unstable, environments. The ERS of a paired male with his primary mate (m), with other females (f) and in total (dashed line) in stable environments (rivers and permanent lakes) is shown in C; and in variable habitats (ephemeral wetlands) in D, in relation to the degree of primary mate attendance. Note that the trade-off between defending his paternity (mate attendance) and securing other copulations favours an attentive strategy in stable and an unattentive one in unreliable environments. From Gauthier (1988).

Gauthier (1988) produced a model that predicts which conditions best favour the occurrences of forced copulations. The pay-off for the male varies in different habitats as the female's nest success varies. For example, Pintails nest in habitats that could change unexpectedly by flooding or by drying out leaving little food for brood rearing, whereas Torrent Ducks nest and rear their offspring on fast-flowing rivers that have little chance of changing. Therefore, the chances of nest success for female Pintails are more variable than for female Torrent Ducks. The second assumption in the model is that female success increases the more a male protects her or a territory where she can feed; male Pintails show no defence of feeding areas and male Torrent Ducks are extremely territorial. The model predicts that the males in stable habitats where females have a high chance of nest success will tend to remain with the female and/or territory (Figure 4.3). On the other hand, in habitats where the risk of total nest failure is high, the chances of a male fathering some offspring will increase if he leaves his primary female and inseminates others.

4.1.4 *Co-operative strategies*

Helper strategies are normally part of co-operative breeding systems when one individual outside the pair forgoes its own breeding attempt in order to assist another individual to reproduce. In birds this usually refers to older offspring helping parents feed younger siblings, helping at the nest (Brown 1987). In precocial waterfowl helping in this sense is not possible and there are rather few examples of co-operative breeding. Magpie Geese in Australia breed in polygamous trios where both females lay in the same nest and incubate and rear the young together with the male (Frith and Davies 1961). We also know of a family of Coscoroba Swans in captivity, where older cygnets assisted by incubating their parents' second clutch and subsequently formed a large brood with their young siblings.

4.1.5 *The true parasite—the Black-headed Duck*

The Black-headed Duck is the only waterfowl species which is a true and apparently obligate parasite. This South American species lays its eggs (usually singly) in the nests of a number of bird species, including those of birds of prey, but the preferred hosts are coots that nest in the duck's marshy habitat (Weller 1968). Studies in captivity indicated that the newly hatched duckling was well adapted to this strategy (Rees and Hillgarth

1984). One duckling, hatching under a Rosy-billed Pochard, before the host's own eggs had hatched, left the nest to fend for itself the day after hatching. Another which hatched with the eggs of the host, stayed with the brood for the first day but progressively distanced itself during the second and was on its own on the third day. This flexibility of post-hatching behaviour clearly enhances the duckling's chances of survival.

4.2 Family behaviour

In nearly all ducks, the female only leads the brood, and there is no family life beyond fledging. Even in those few species, such as a number of riverine species, where both sexes lead the brood, family life finishes with fledging and dispersal of the young. Whistling ducks and sheldgeese do show an extended period of parental care but it is in the geese and swans that family behaviour is best developed and studied in most detail.

Northern swans remain territorial until the young are fledged or later, but most goose families gather in loose flocks soon after hatching. During the pre-fledging period, Pink-footed Geese invest in parental care by reducing competition for the young by keeping other families away from the immediate vicinity. The parents protect their young from predators by visual scanning (mainly by the male), by seeking proximity with other geese whilst resting, and, if necessary, by active defence of the brood. The cost of this investment is that parents sleep and preen less than non-parental pairs. Because of their larger investment, parental males have less time to feed than females but compensate by pecking more rapidly (Lazarus and Inglis (1978).

This parental investment continues after fledging and after migration to the wintering grounds, where all species gather in flocks. Within the family and the flock, geese associate together according to the degree of trade-off or conflict between the disadvantages of competing for the same food supply and the feeding and anti-predator advantages of association.

Family associations in swans provide an excellent and well-studied example of parent–offspring interaction (Scott 1980). All families remain together throughout winter and depart for the breeding grounds together. The proximity of young to the adults' nesting area during the breeding season is unknown, but a substantial proportion (68%) of swans associate with parents in their second winter. Yearlings associate both with parents without young of the year, and those with cygnets, forming 'super-families', consisting of parents, cygnets and yearlings. Some young are known to associate with their parents in their third, fourth and even fifth winter (Evans 1979).

Large Canada Geese also show remarkable family cohesiveness; families were intact on 96% of occasions in winter, many stayed together through the spring, and 15% of yearlings also associated with their parents (Raveling 1969). The rule for *Anser* geese is for families together through winter and spring; 75% of Lesser Snow Goose families were intact soon after arrival on the breeding grounds and some associated the next winter (Prevett and MacInnes 1980).

Evidence indicates that family stability in smaller species of geese depends on the situation. Jones and Jones (1966) noted that families of Black Brant broke up before autumn migration, but this was in large concentrations in a very disturbed area. In other studies on Brent, most birds are in families through the winter (e.g. O'Briain 1987). Johnson and Raveling (1988) found that only 5% of juvenile Cackling Geese were in families and 5% of pairs together in winter flocks. They suggested that the weaker family bonds in small geese is a result of greater gregariousness brought about by greater predator pressure and as an adaptation for the grazing habit (large flocks are better at managing vegetation to maintain its nutritional quality—see page 19). The Barnacle Goose is larger than the Cackling Goose and closely related. It does show a high degree of family cohesiveness but break-up of families begins in midwinter and only a third of goslings are in families when the birds make the final flight to the breeding grounds (Black and Owen 1989a).

There certainly seems to be a trend of increasing cohesiveness with increasing body size in geese and swans. The delay of maturity and achievement of full adult size and weight also increases with body size. For example, yearling Bewick's Swans are significantly lighter and by some measurements smaller than adults in their second winter (Evans and Kear 1978), whereas Barnacle Geese achieve adult size in the first autumn and adult weight in their second summer (Owen and Ogilvie 1979). Small geese are more vulnerable to predation in winter than large ones and gather in dense flocks, where the maintenance of family cohesion is difficult. There must be a trade-off between the benefits of family life and the costs in terms of searching time and energy in maintaining the family unit. In the Barnacle Goose, family cohesion has become less strong as numbers in the population have increased.

The suggestion of Johnson and Raveling (1988) has some validity, but it requires further testing. We suggest that the most powerful determinant of family cohesiveness is the size of the population and of the wintering group and the degree of disturbance it is exposed to. Group size and degree of disturbance are not, however, independent of body size, for the reasons outlined by Johnson and Raveling. The balance of costs and benefits of maintaining the family bond shifts as the size of the group increases. This suggestion must be examined within a species, by testing the prediction that families are less cohesive in larger and more volatile groups.

To be maintained by natural selection, the long-term family bond must confer survival advantages to the young or increase parental survival or breeding success. In Bewick's Swans Scott (1980) demonstrated that the young of dominant families had greater feeding opportunities than subordinate groups, and that this implied greater survival. Family groups dominated smaller units, so association with offspring benefited the whole group. Family members benefited from being in the group whether or not the whole group was involved in interactions.

Boyd (1953) found that the success of individual White-fronted Geese in encounters with others depends on the size of their group (Figure 4.4a). Pairs beat single individuals, families beat pairs and family success rate increases with size. This relationship could come about in several ways. The degree of aggression shown by an individual may increase with group size and success increase with aggression, or there may be direct or indirect (as

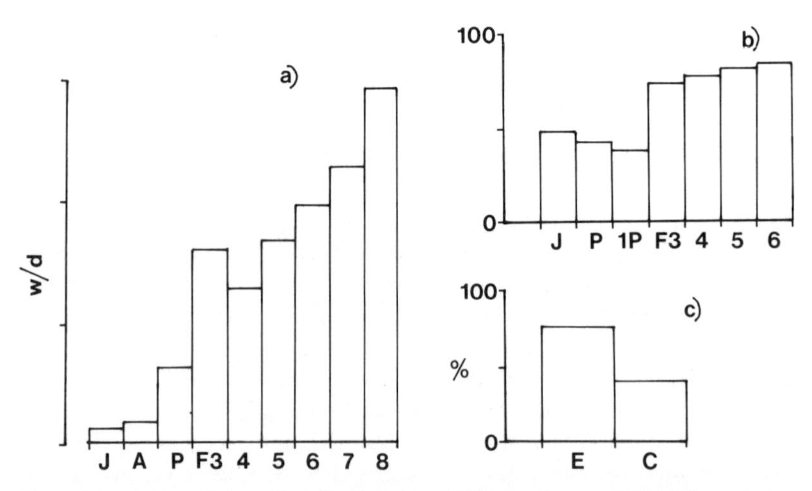

Figure 4.4 a) The success rate (wins/defeats) of different classes of White-fronted Geese against conspecifics of all other classes. Derived from Boyd (1953) and reproduced from Owen (1980a); b) the percentage of occasions that individual Barnacle Geese of different classes are found at the leading edge of the flock; and c) the difference in biomass between the edge and centre of fields, averaged for a whole season. b) and c) from Black and Owen 1989b.

for swans) assistance by group members in conflicts. Detailed studies on Barnacle Geese demonstrated that group size was the most important predictor of success; neither body size nor age had a significant effect (Black and Owen 1989b). Individuals adjusted their aggressiveness to the rank of their opponents such that they used the least effort to win an encounter. Families were predominantly found at the leading edge of the flock (Figure 4.4b), where the green biomass was higher (Figure 4.4c). There is, as yet, no direct evidence that being in a family confers future selective advantages for young geese, though there were clear feeding advantages for attached goslings and they grew fatter than lone juveniles (Black and Owen 1989a). In Snow Geese, both goslings and adults in families survived significantly better than non-family birds, but this was in a situation where mortality was artificial, from shooting (Prevett and MacInnes 1980). Evidence is accumulating, however, on the survival advantage of extended parental care for young Anserinae.

Are there costs for the parents in terms of future productivity? Black and Owen (1989a) found that association with goslings did not reduce the chances of surviving to, or breeding, in the following season. Indeed, pairs which associated with young for longest tended to be most successful. We suggested that this was because young that stayed in the family contributed to the success of the group.

The balance of advantage in family cohesion changes over time and the strongest of bonds break when competition between family members outweighs the advantage of group maintenance. Black and Owen (1989b) examined the process of family break-up in detail in Barnacle Geese (Figure 4.5). Nearly all families were intact when they arrived from the breeding grounds and there was little separation until the new year. The rate of separation increased on the first stage of the migration north, and during the staging period; only a third of goslings were in families when the birds left for the breeding grounds (Figure 4.5a). As the winter progresses, parents increasingly threaten their young, and the appeasing greeting displays of goslings increase in parallel until May, when most families are separated (Figure 4.5b).

Family break-up occurs through brood reduction rather than through some groups remaining intact and others disbanding. In view of this, as might be expected, the duration of association increased with brood size. Black and Owen (1989a) hypothesised that the goslings that remain in the family assisted their parents by being aggressive towards competitors to the family (including their subordinate siblings) and decreasing the parental investment in group defence. There is some unpublished evidence from

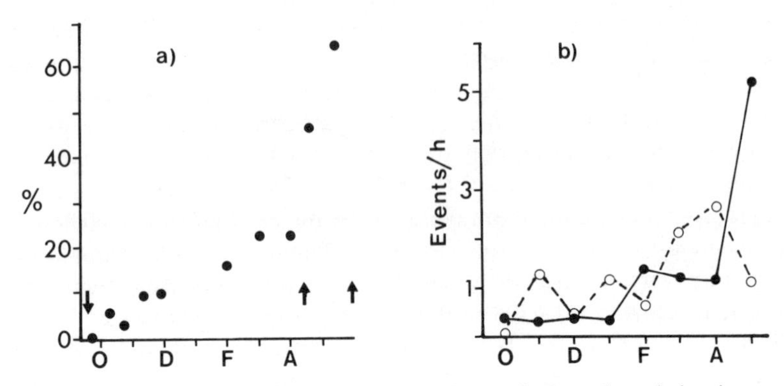

Figure 4.5 a) The pattern of break-up in families of Barnacle Geese through the winter and spring (arrows indicate migration times). b) The frequency of parental threats and attacks at their own offspring (both sexes of parent combined—closed circles and the frequency of the goslings' appeasement greeting displays—open circles). From Black and Owen (1989b).

captivity and from the wild that those young Barnacle Geese that leave the family last are the larger, more aggressive goslings and that they do contribute to their parents' attacks on competitors as well as assisting with the family vigilance burden.

The size of the benefit to families is extremely difficult to quantify under natural conditions, but it is clear that there are trade-offs between the benefits and costs of cohesion and that the range of family loyalty found in waterfowl depends not only on the species, but also on the situations in which they find themselves.

4.3 Territoriality

Animals forage either in groups or alone, and the two methods have different advantages according to the balance between the benefits of group foraging—food finding and minimising individual predation risk—and the costs in relation to competition for local resources (Pulliam and Caraco 1984). Many waterfowl are territorial during the breeding season, but few maintain territories in winter.

Year-round territoriality is particularly characteristic of riverine species, and there has been parallel evolution in unrelated groups. The Blue Duck of New Zealand defends territories which includes rapids in which it feeds, leaving undefended other stretches of river less suitable for feeding. The territories, at least at critical times of the year, are closely matched to the food supply (Eldridge 1986b). Territorial defence displays and foraging are concentrated in the morning and evening, but non-territorial birds are allowed to feed on territories at midday. Since invertebrate availability peaked morning and evening, the effort expended in territorial defence was also related to the food supply.

The unrelated African Black Duck, also a riverine species, effectively excludes conspecifics from 90% of its territory. Non-territorial birds are excluded from the most productive riverine habitat and do not breed (Ball *et al.* 1978). Other river specialists share many characteristics with the above two examples (Kear and Steel 1971). The Salvadori's Duck of New Guinea and the Torrent Ducks of South America, are similarly territorial year-round. The pair-bond is strong and long lasting in the Torrent (Eldridge 1986a), African Black (McKinney *et al.* 1978) and Salvadori's Duck (Kear 1975). Although the Blue Duck's pair-bond is of shorter duration, both pair members assist in defence of the territory (Eldridge 1986b). All four species have well-developed spurs at the angle of the wing, which are used in fighting.

The two flightless species of Steamer Ducks (and the flying species—mentioned later) of South American also defend year-round territories along the shore and have long-term pair-bonds, features that are related to the maintenance of a secure food supply under harsh climatic regimes (Weller 1976). In only one species of waterfowl has it been demonstrated that pairs defend different territories in summer and winter—the Barrow's Goldeneye (Savard 1988). The female rears the young alone on fast-flowing rivers whilst the male retires to the coast to moult. The pairs reunite during winter and defend a territory along a stretch of shore. The function of the winter territory is unclear, but the defence of a food supply for the pair seems the most likely explanation.

The significance of territoriality in these species (apart perhaps for the Goldeneye) is clearly food-related; only birds defending a good food supply are able to breed. There are costs in terms of increased individual predation risks, to a territorial lifestyle, but these have not been quantified for any waterfowl species.

4.4 Flock behaviour

Most waterfowl are gregarious in the non-breeding season, and their gregariousness is not simply related to concentration in good food patches or safe resting places. Even when the environment is not patchy, many species gather in flocks. There are two, not necessarily conflicting, reasons as to why birds gather in flocks: the anti-predator benefit (improved detection and reduced individual risk in larger groups), and the advantage in location and exploitation of food (Pulliam and Millikan 1982).

In White-fronted Geese, Lazarus (1978) examined two hypotheses related to the anti-predator function of flock size. The first predicted that the larger the flock, the smaller the proportion of birds vigilant, since for an equal degree of predator detection, each flock member has to be vigilant for less time. This was indeed the case, as shown in Figure 4.6a. Having found a similar relationship in Barnacle Geese, Drent and Swierstra (1977) suggested that the reason for this relationship was purely because of an 'edge effect': peripheral individuals were more vigilant than central ones and the relative length of perimeter declines with flock size. Lazarus modelled the relationship by using differential vigilance values for central and edge birds, as shown in Figure 4.6b. The relationship for White-fronts is very similar to that predicted by the model when the proportion of birds vigilant in the edge of the flock $Q = 1$ and the proportion of alert birds in the centre $q = 0$ (not a realistic relationship). Further work on the 'edge effect' on Brent Geese (Inglis and Lazarus 1981) found a relationship similar to that described when $Q = 0.7$ and $q = 0.3$. They concluded that the pattern of vigilance found in goose flocks could be explained by the edge effect.

Why should birds on flock edges be more vigilant? Lazarus (1978) examined the concept of 'domain of danger' proposed by Hamilton (1971); the domain is the area around an individual within which a predator would be closer to the bird than to any other flock member. The prediction is that the larger the domain of danger, the more time an individual should devote to predator scanning. This was tested on White-fronted Geese by measuring the vigilance of birds with different domains and the results supported the hypothesis. These lines of evidence suggest an anti-predator advantage to flocking, where the time spent in scanning decreases with flock size, the relationship being brought about by individuals modifying their vigilant behaviour in relation to their potential vulnerability.

The anti-predator function of flocking was examined by Poysa (1987a) in Teal, by testing two predictions: that flock size should increase with increasing predation risk, and that, when a predator was present and Teal were aware of it, the proportion of time-scanning should decrease with group size. Neither of the predictions matched the field observations; rather, the behaviour of the Teal was more likely to be explained by their hunger and the conflicting pressure of feeding competition with increasing

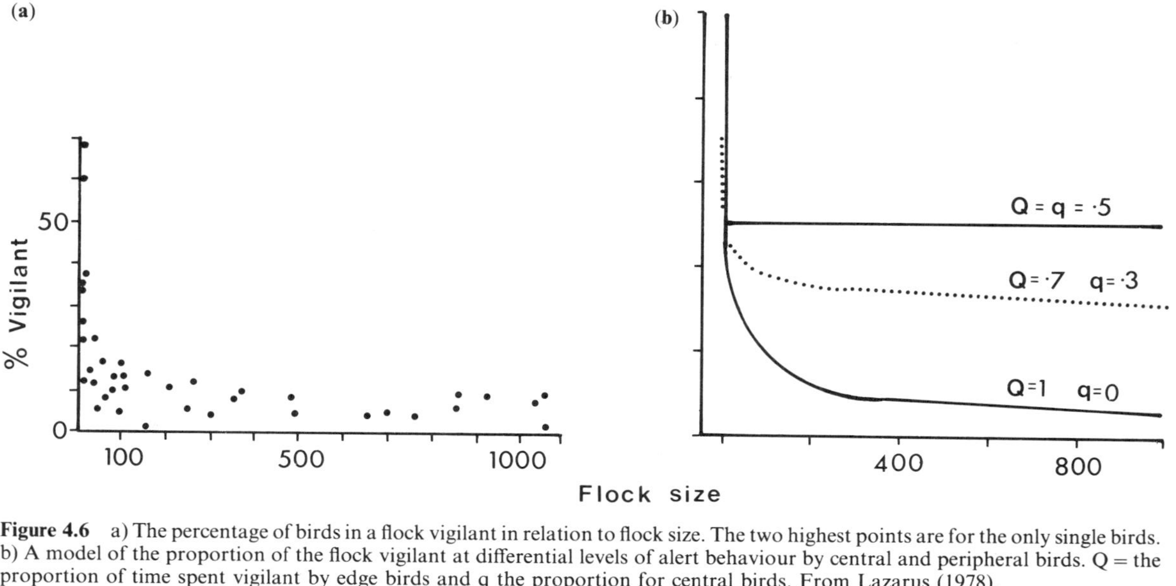

Figure 4.6 a) The percentage of birds in a flock vigilant in relation to flock size. The two highest points are for the only single birds. b) A model of the proportion of the flock vigilant at differential levels of alert behaviour by central and peripheral birds. Q = the proportion of time spent vigilant by edge birds and q the proportion for central birds. From Lazarus (1978).

flock size. Further studies (Poysa 1987b) indicated that in Teal the relationship between the costs and benefits of flocking favoured the hypothesis that flocking is an adaptation to the exploitation of patchy food sources than an anti-predator mechanism. Poysa presented evidence that: a) flock size was related to the abundance of food; b) feeding flocks attracted flying birds; and c) arriving birds landed close to those already there and copied their feeding methods. Predictions of the anti-predator hypothesis were not supported in the studies of Teal.

Experimental evidence for the feeding advantage of flocking was provided by Drent and Swierstra (1977). They placed model geese, some in the grazing posture and some in the alert (head-up) posture. Flocks of model geese were successful in attracting flying flocks and causing them to land. Furthermore, flocks containing a higher proportion of 'grazing' models were more effective at attracting birds to land and stay longer in the vicinity of models than were mostly alert models. In one trial, the number of goose-hours spent feeding in the vicinity (a combination of number of flocks landing, flock size and duration of stay) of a flock of 25 grazers and 5 alert models was 637 goose-hours, compared with only 34 goose-hours for a flock of the same size containing equal proportions of feeding and alert models. Drent and Swierstra suggested that the 'alert' posture was a general scanning of the environment, not only for predators, and that birds flying over a feeding flock used the proportion of alert birds to assess the potential quality of the feeding situation.

Another advantage of gregariousness that is also concerned with food finding was first proposed by Ward and Zahavi (1973); they suggested that assemblages, mainly roosts, acted as 'information centres'. Birds which had experienced poor feeding conditions previously somehow identified more successful feeders and followed them to their feeding areas. Some aspect of behaviour gave clues, possibly the earliness of the departure of experienced birds or some other behavioural trait.

There are two studies that address the predictions arising from this hypothesis in waterfowl. Barnacle Geese in the Netherlands gather in post-roost flocks in the morning before dispersing to diverse feeding grounds (Ydenberg *et al.* 1983). They found that, as shown in Figure 4.7a, the duration of the post-roost gathering was related to the lateness of the flight to roost the previous evening. Late departures are related to cold temperatures and large flock sizes, both of which, it is argued, are associated with low food availability. Ydenberg *et al.* propose that on experience of poor feeding conditions on the previous day (late departure for roost), birds must give more consideration to where to go the following

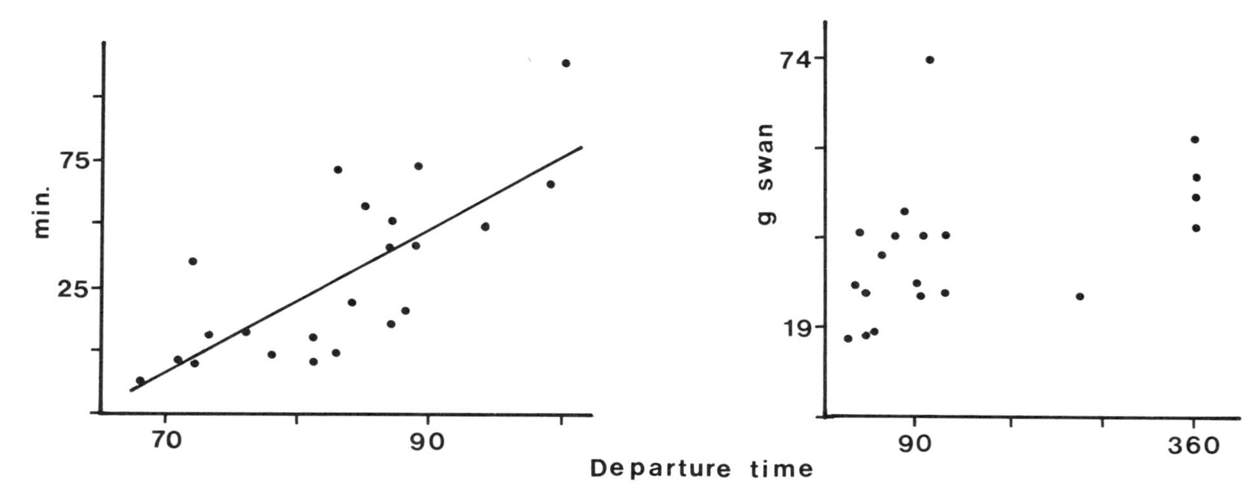

Figure 4.7 Left, the duration of the post-roost gathering period of Barnacle Geese (in minutes) in relation to the lateness of departure from the feeding grounds the previous evening (the proportion of daylight that has gone). From Ydenberg *et al.* 1983. Right, the time between the first departure of Whooper Swans from the feeding area, a provisioned food supply, in relation to the amount of food available per swan (from Black 1988).

day, hence the increased duration of the post-roost gathering. Their evidence is suggestive but not convincing; birds could, for example, be reacting to current temperatures, which are correlated with those the previous night, by delaying departure to unprofitable feeding grounds (frost often disappears from foliage later in the day and even birds already on the feeding grounds on frosty mornings feed little until this happens). Another possibility is that the hunger drive of birds which had fed late the previous night was less than for earlier departures and that this delayed their flight from the post-roost gathering the following day.

A clear mechanism for the use of visual signalling in food finding is the pre-flight signalling behaviour (head-pumping) of Whooper and Bewick's Swans (Black 1988). Those individuals which were about to depart to a predictable and rich feeding area (a provisioned site), carried out fewer pre-flight signals and departed sooner after indicating their intention, than those which were about to depart to poorer and less predictable feeding areas (grass fields). The departure time from the good area was significantly related to the amount of food available (birds exhausted the food supply—Figure 4.7b). However, some groups, notably families, were dominant and subordinates obtained less food. The first leavers were significantly more likely to be subordinate (hungry) birds and they initiated pre-flight signals in such a way that attracted others to depart with them and form a new flock. This study provides no direct evidence that social signalling is used in food finding, but does illustrate a way in which the birds could monitor each other's behaviour to obtain information about food.

4.5 Interspecific interactions

It is a long-established principle that animals with closely matched requirements for food and other resources compete with each other, in the extreme, to the exclusion of one competitor. If closely related species overlap in range they must develop specialisations of behaviour or morphological adaptations which minimise the competition between them (see Chapter 2). In situations where food was not previously limiting, changes in the abundance of one animal may bring it into competition with another. Interspecific conflict is not often observed, or detected, but a number of direct and indirect examples are available for waterfowl.

Barnacle and Pink-footed Geese moult in the same areas of east Greenland, sometimes in single species and sometimes in mixed flocks. The numbers of both species have increased in recent decades, because of factors operating on the wintering grounds (Owen *et al.* 1986). In Greenland there is evidence that the increased numbers have reached the carrying capacity of the habitat and there is competition between them (Madsen and Mortensen 1987). Both species in isolation feed on sedges and grasses, but in mixed-species flocks the amount of moss (an inferior food) in the diet increases, especially in the Barnacle Goose. Both species spend 41–46% of the day feeding when alone, but the Barnacle Goose spends 62% feeding on mixed moulting sites, whereas the time budget of the larger Pink-foot is unaffected. Madsen and Mortensen did not see evidence of active interspecific aggression between the species when both were present. They suggest that the Barnacles will learn to avoid areas where they fare less well and to occupy alternative moulting grounds in future years. Competition between the species, however, may eventually lead to lower survival of the poor competitor as it is increasingly driven into suboptimal habitats.

Interspecific competition in the wintering grounds, leading to the displacement of one species, has also been described (Madsen 1985). Greylags occupied an area of farmland in Denmark in early autumn, before Pink-feet arrived from their breeding grounds in October. When the migrants arrived, the feeding distribution of the Greylag changed, as Pink-feet, which occurred in larger numbers, occupied the best feeding sites on stubble fields—the preferred feeding habitat of the Greylag.

The Barrow's Goldeneye is territorial during the breeding season, and directs most of its efforts in excluding conspecifics. The species does, however, direct aggression towards other species, and Savard and Smith (1987) present evidence to suggest that there are two different bases for interspecific attacks, which were directed at species which have a) similar diets and b) similar plumages, to the Barrow's Goldeneye. They argue that aggression against dabbling ducks is misdirected because they are not competitors but happen to show similar plumage characteristics to the attacker.

The most violent interspecific aggression yet described is in the Flying Steamer Duck, a species which is highly territorial throughout the year. Not only do the drakes defend territories against their own species, but they stalk and attack flocks and individuals of other species, including grebes and dabbling ducks. Neuchterlein and Storer (1985) witnessed male Steamer Ducks beating and killing flightless Red Shovelers and found a number of carcasses of the same species with evidence of death by beating. The species attacked were not close competitors with the Steamer Ducks for food, although competition for food could not be excluded as a reason for the aggression. It was suggested that the risks of such attacks to the Steamer Ducks were minimal, because of their adaptations for fighting, and that a possible benefit of the 'public beatings' was to demonstrate fighting abilities to the female and reinforce the pair-bond.

Waterfowl demonstrate a wide range of social behaviour, and in general all aspects are ultimately related to individual fitness through their effect on survival and breeding performance, through affecting the ways in which individuals can exploit the food supply and other resources in competition with others.

CHAPTER FIVE

MOVEMENTS AND MIGRATIONS

Bird migration has always fascinated Man and a great deal of effort has gone into researching the where, why and how of long distance movements. Because waterfowl provide a valuable resource, the group has received as much attention as any family of birds. Most of the studies have been in North America, as the classic works of Hochbaum (1955) and Bellrose (1968) testify.

Movements of animals occur in response to changes in the availability of food or some other essential resource, or when their physiological requirements dictate the need for different habitats or foods. If the changes are predictable, movements tend to be regular and these are the movements commonly perceived as migrations (see Baker 1978 for review). They are normally over long distances, and usually along an altitudinal or climatological gradient. Movements in response to unpredictable events, as often occur in tropical areas, are opportunistic and the species tend to be nomadic. This section also deals with movements over shorter distances, within breeding or wintering areas, with range extensions and with loyalty to particular locations. A special form of migration—the moult migration—is dealt with separately.

5.1 Monitoring migration

Most of the information on waterfowl migration has been the result of marking birds with unique identifiers and recovering them later. Metal leg rings are by far the most common method of marking, but not the most efficient, since subsequent information depends on retrieving the ring through recapture or finding the bird dead. Biologists studying waterfowl do have an important advantage over most, in that dead birds are usually recovered, since most deaths are brought about deliberately by Man. Recovery rates of waterfowl are in the region of 15–25% of those ringed—a very high proportion compared with most birds.

Ringing and recovery however, contribute a minimum of information:

the location of a bird on two occasions. A large number of such observations are needed to delineate migration routes and to describe the chronology of movements. New individual marking schemes have more potential, especially those that provide a number of records of living individuals that do not depend on the bird being in the hand. The simplest is a larger version of a leg ring with large characters that can be read at a distance in the field (Figure 5.1). The longer necked species such as many geese and swans can take neck-collars without harm and these give the opportunity of larger characters and are much easier to see when the birds are on water or in tall vegetation. Such marking can be used in combination with plumage dyes which draw attention to the marked birds.

Radiotelemetry has been used extensively to study short-term movements and the recent development of satellite tracking systems provides enormous potential for monitoring migration in great detail and over a very wide range. As yet the radio packs for satellite tracking are rather large and have only been used successfully on swans. Further miniaturisation will, however, eventually extend the range of species, at least to the larger geese.

Migrating birds are sometimes seen, and opportunistic observations often make substantial contributions to knowledge of the patterns of

Figure 5.1 An upending Bewick's Swan carrying an engraved plastic ring with characters that can be read with a telescope at distances of up to 300 m. The use of such rings has been invaluable in population studies of geese and swans.

movements, especially where observatories are located near narrow migration corridors or on regular routes. In the USA, there have been co-ordinated efforts to make observations along complete flyways. Flights can also be followed in light aircraft which can track migrating flocks.

Radar, in combination with visual sightings, has been used extensively to monitor migration; it gives information on the altitudes of migrating birds as well as their routes and timing. It is also one of the few methods that can be used at night, although other information is needed if species are to be identified. Sea ducks are involved in some of the most spectacular migrations, because they are so numerous and because they are loath to fly overland and are often funnelled through narrow straits and across peninsulas. The traditional nature of these movements is used by native peoples to shoot large numbers of Eiders (Bellrose 1978). The massive passage of Eiders through the western part of the Baltic in spring is spectacular, and has been studied by a combination of radar and visual observations. In a matter of days about 800 000 Eiders migrate northwards in spring and 290 000 fly along the narrow strait between the island of Oland and mainland Sweden (Alerstam *et al.* 1974).

5.2 Movement patterns

Migrations enable mobile animals to exploit periodically or seasonally available resources. For example, terrestrial birds breeding in the high Arctic cannot remain there over winter. The same birds are unable to breed on their wintering ranges because the conditions are not suitable, or because those niches are occupied by species better adapted to a more temperate environment. Among waterfowl there is a wide variety of movement patterns between and within species, which are adapted to prevailing environmental conditions.

5.2.1 *Arctic breeders*

The typical waterfowl migrations are of birds breeding in the Arctic or north temperate areas and wintering in temperate zones. Species which travel over continental land masses have the opportunity of stopping at intervals to rest and refuel. The well-defined migration corridors for waterfowl in North America have been investigated in detail (Bellrose 1968). Ducks breeding in the prairie regions of Canada and the northern

United States gather in August and September around larger water bodies before migrating southwards. The largest portion move down the Mississippi Valley, stopping on the way at traditional floodlands and pools, moving ahead of the advancing winter. Most end up on the vast marshes of the Gulf coast, but some species such as Mallard move only as far as they have to to find food. If the winter is mild, a high proportion stay in northerly states, as close as possible to the breeding grounds. Other flyways follow the coasts, from Alaska south to California and Mexico and from eastern Canada to the coasts of the Carolinas south to Florida.

The routes followed are highly traditional and the staging areas used year after year. The precision of the routes is illustrated by the three

Figure 5.2 The migration routes (arrows), staging areas (circled) and breeding grounds (shaded black) of the three populations of Barnacle Geese which winter in western Europe and breed in different parts of the North Atlantic. Reproduced from Owen (1980a).

populations of Barnacle Geese wintering in Europe (Figure 5.2). Despite the fact that their wintering areas are very close together, the three populations are discrete. Those breeding in Greenland stop in south-east Iceland on their way south and winter in western Ireland and Scotland. The small population breeding in the Svalbard archipelago stages in the most southerly of the islands before making the 3000 km flight over the sea to wintering grounds on the Solway Firth. The Russian breeders travel through the Baltic, stopping on the coasts of Germany and Friesland before proceeding to wintering grounds in the west and south of the Netherlands (Owen 1980a).

In spring each population leaves its wintering grounds in the latter half of April to highly traditional staging areas. The Greenland stock stops for two or three weeks in the valleys of north-west Iceland before proceeding to the breeding grounds. At exactly the same time of year, the Svalbard geese are staging on islands off the Norwegian coast just south of the Arctic Circle, and the Siberian breeders are on the Swedish island of Gotland and parts of Estonia. The timing of their migrations is so precise that mass movements occur within the same few days each year (see e.g. Owen and Gullestad 1984).

5.2.2 *Nomadic species*

The seasonal movements of tropical species are related to the occurrence of rains and the temporary flooding of wetlands. The Australian Grey Teal is a nomadic species, whose movements are described as 'explosive random dispersals' (Frith 1982). Extensive floods in inland Australia provide suitable conditions for breeding, and when the waters recede, the ducks move out to find new habitats. One such movement followed a flood in the Murray–Darling Basin in the inland south-east in 1956. In the following year, Grey Teal scattered in all directions and ringing studies indicated that many had reached the coast and some had dispersed as far as New Guinea in the north and New Zealand to the east (Frith 1962). Many other species which use temporary tropical and subtropical wetlands show similar capacity for opportunism and irruptive dispersal.

5.2.3 *Changes in range*

Waterfowl undergo expansions and contractions of range as climate or other conditions change. In historical times some major changes have come about in the range of some North American species. The breeding distribution of the Lesser Snow Goose has extended southwards and westwards from its traditional strongholds in northern Hudson Bay and the Foxe Basin (Kerbes 1975). The large colony at McConnell River on the west side of Hudson Bay was established in the 1940s, when the breeding range was moving southwards. The reasons for these major shifts are unknown, but the species nests in large colonies, which go through periods of expansion and contraction. The colony at La Perouse Bay expanded in the 1970s and early 1980s to the extent that the capacity of the area was exceeded and reproductive success declined (Cooch *et al.* 1989). The changes in winter distribution, brought about by the availability of food on agricultural land have probably also had an impact on the nesting distribution of Snow Geese.

The Gadwall has also extended its range in Europe and North America in recent decades. The eastward expansion in the USA is thought to be due to the creation of habitat in the form of impoundments and the creation of permanent waters on refuges. There were also attempts to boost local populations by importing birds from the west, but these were not a great success (Henny and Holgerson 1974). In Britain releases of Gadwall in the 1960s and 1970s did lead to the establishment of substantial numbers breeding, but the main increases also corresponded with the northward and westward range expansion in continental Europe, where many of Britain's wintering birds originate (Fox and Mitchell 1988). The increases in Europe may also be due to the creation of new freshwaters, which the Gadwall has discovered, in the lowlands.

5.3 The moult migration

In common with a number of other groups of mainly water birds, waterfowl moult their primary and secondary flight feathers simultaneously and are flightless until these have regrown. The need for safety is paramount at this time and many species make a special migration after breeding to the safety of moulting grounds (Salomonsen 1968). The flightless moult is dealt with in more detail in Chapter 3.

Moult migration is not well developed in swans; both parents care for the young and moult near the nesting area. Non-breeders and immatures gather in flocks to moult but only in the Mute Swan has substantial movement been reported. Non-breeders from the large population of Mute Swans in the western Baltic and parts of southern Sweden gather to moult in shallow bays on the coast of south-west Sweden (Mathiasson 1973). The birds come from Danish and Swedish parts of the Baltic and also from the Swedish lake populations to the north.

Immature and non-breeding birds of several populations of geese have a well-defined moult migration, and this is almost invariably to the north of the breeding birds, which stay with their young. Two classic cases are the migration of Pink-footed Geese from Iceland to Greenland moulting areas in early July (Taylor 1953), and the journey of at least 1600 km made by immature Canada Geese from the United States to Canada (Stirling and Dzubin 1967). The recently developed moult movements of the Greylag and the Canada Goose in Europe provide the only known exceptions to the northerly trend (Figure 5.3). Salomonsen (1968) gave several possible

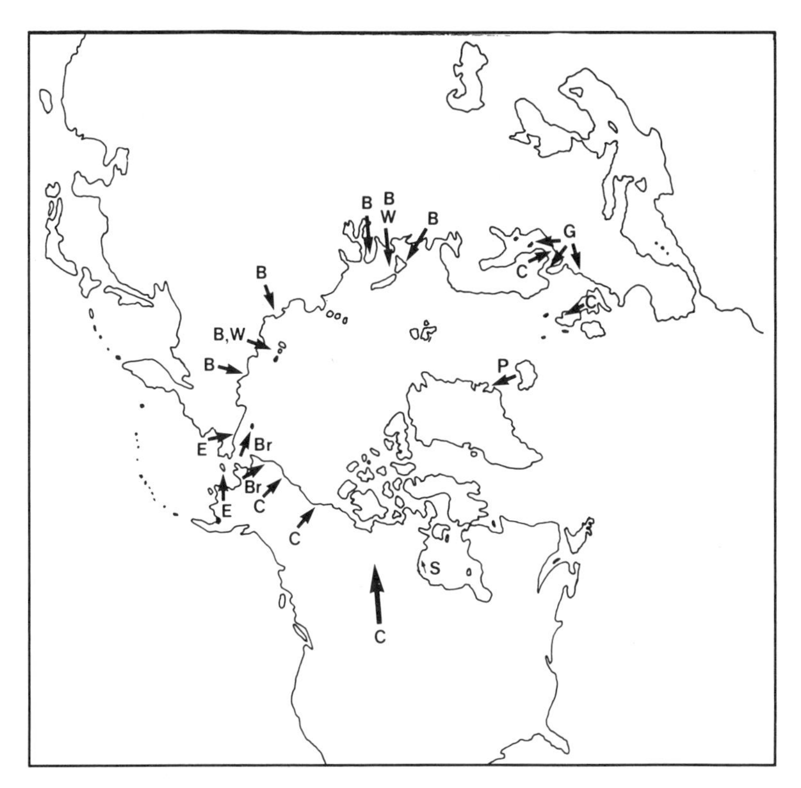

Figure 5.3 The moult migrations of geese. Abbreviations as follows: B—Bean; Br—Brent; C—Canada; E—Emperor; G—Greylag; P—Pink-footed; S—Snow; W—White-fronted. updated from Salomonsen (1968), reproduced from Owen (1980a). Note that the only exception to the rule of northerly movement is that of introduced Canada Geese in the Baltic.

reasons for the northerly direction of movement, including a prolonged spring migration, and the avoidance of competition with breeding birds. Stirling and Dzubin (1967) proposed that migrating birds profited from greater safety as well as lack of feeding competition, and Ebbinge and Ebbinge-Dallmeijer (1977) suggested that the birds moved northwards in search for longer days which would enable them to detect predators for a greater proportion of the time (the 'avoidance of the dangerous dark' hypothesis).

None of these theories explains all the known migrations and Owen and Ogilvie (1979) proposed an alternative explanation. Their hypothesis was that the birds were moving to areas where there was more nutritious food (effectively capitalising on another flush of growth). This is supported by the fact that species such as the Lesser White-fronted Goose (Ekman 1922 quoted in Salomonsen 1968) and the Greenland White-front move up in altitude to moult. This movement up a climatological gradient is the equivalent of flying north. The migrations which are southerly, such as that of Greylag Geese into the Netherlands, are to areas with particularly favourable feeding conditions (Van Eerden 1990).

One of the most celebrated and clear moult migrations is that of the Shelduck in Europe. Adult ducks leave their half-grown young on the breeding grounds and migrate, together with non-breeding and immature birds, to the German Waddensee, where 200 000 or more gather on the vast sandflats to moult. The birds originate from Britain, Scandinavia and even southern France (Salomonsen 1968). A few smaller groups occur, notably on several estuaries in Britain such as Bridgwater Bay on the Severn, and the Dee (Owen *et al.* 1986), but these are of negligible importance compared to those in the Waddensee.

In the dabbling and diving ducks, the males leave the females soon after incubation begins and males of several species gather together in massive moulting aggregations, usually on large bodies of water. More than 10 000 ducks of eight species moult on Takslesluk Lake in western Alaska. Recoveries of birds ringed there were from a very wide area, from central Siberia to the far eastern United States, and as far south as Mexico (King 1973). Similar mass gathering grounds for dabbling ducks occur in southern USSR; an estimated 380 000 ducks moult on the Volga Delta on the north side of the Caspian Sea (Krivonosov 1970). At this time males lose their colourful breeding plumage and assume a cryptic, female-like 'eclipse' plumage which makes them less vulnerable to predation.

All species of sea ducks show well-marked moult migrations, of which one of the most notable is the movement of male and non-breeding King

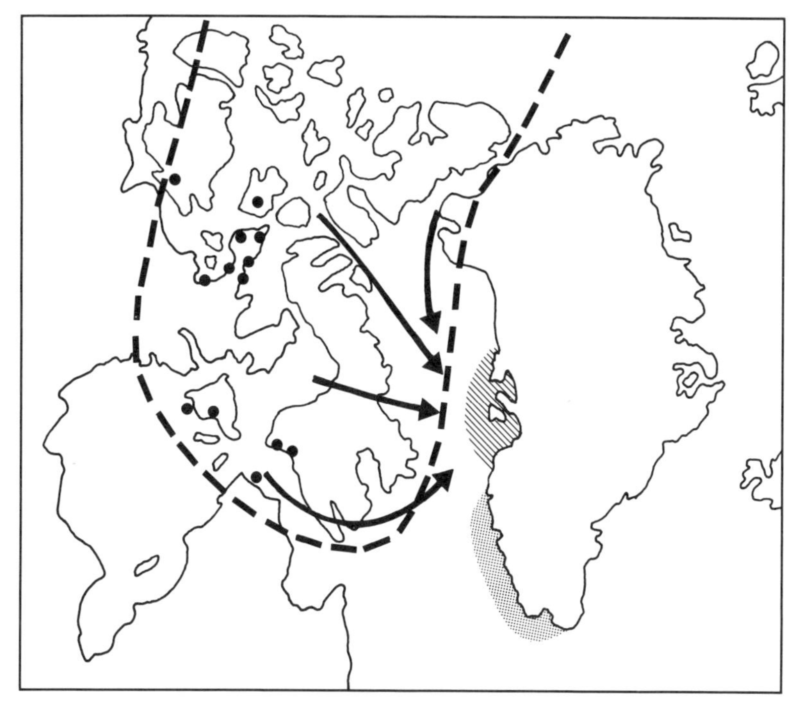

Figure 5.4 The moult migration of the King Eider to the moulting area around Disko Bay in west Greenland (hatched). The dashed line indicates the breeding range and the dots indicate recoveries within the breeding season of birds ringed as moulters in Greenland, and the arrows the direction of movement. The stippled area in south-west Greenland is the wintering range. From Salomonsen (1968).

Eiders from the breeding areas of north-eastern Canada to the area around Disko Bay in west Greenland (Figure 5.4). Up to 100 000 birds moult there before moving south to the south-west Greenland wintering area in September (Salomonsen 1968). Spectacular concentrations of several species of sea ducks, from breeding grounds in the north and east, gather to moult in the shallow coastal waters of the Baltic, mainly around Denmark. The total number of birds of all species together in this area is not far short of 500 000 (Joensen 1973).

As Salomonsen (1968) pointed out, a characteristic of duck (as opposed to goose) moult migrations is that they are invariably to an area with a milder climate than the breeding zones, in many cases on the way to the winter quarters, or even in the wintering range itself. Those populations which breed in the south show less of a tendency to move to moult, and there is no known moult migration (of substantial distance) in tropical waterfowl. It is clearly advantageous for birds which have no chance of breeding (immatures) or whose reproductive effort is over for the season (males and failed-breeding females) to establish themselves in productive areas to the south, closer to the wintering area. Substantial selective pressures must have been at work for goose populations to develop such long and clear-cut moult migrations; the gaining of a feeding advantage which improves future survival seems the most likely source of selection.

5.4 The timing and mechanics of migration

A number of different factors determine the best time for migration, and birds have a range of clues available which they can use to indicate the 'correct' time. The time available, particularly for Arctic species, for breeding is severely limited, so it is vital that the spring migration is timed so that the birds do not have to spend valuable energy resources waiting for conditions to be favourable or, on the other hand, arriving too late to complete a breeding cycle. Preparation for migration, however, begins much earlier, as days begin to lengthen in spring and the birds begin to lay down reserves in preparation for the demands of the flight and of breeding.

5.4.1 *Photoperiod*

Numerous studies indicate that the changing daylength is a predictable factor which triggers off migration in birds. Evidence on this in waterfowl comes from an accidental 'experiment' at the Wildfowl and Wetlands Trust, which gave a clear demonstration of the daylength stimulus for Bewick's Swans (Rees 1982). Swans and other waterfowl are fed in the Trust's enclosures, in midwinter under floodlights, to provide a public spectacle. The exposure of swans to the lights varied from year to year as management practices changed. The birds' departure patterns under three different lighting regimes are shown in Figure 5.5. The departure dates are clearly different under the three regimes; the time when 50% of the flock had departed was six weeks earlier under the most extended illumination than under natural daylength. The results were in different years, but departure patterns from other sites without floodlighting differed in the same way in the same years, confirming that it was a daylength rather than a year effect. In one year some individually marked swans were identified on migration in Germany, but following a cold spell there they returned

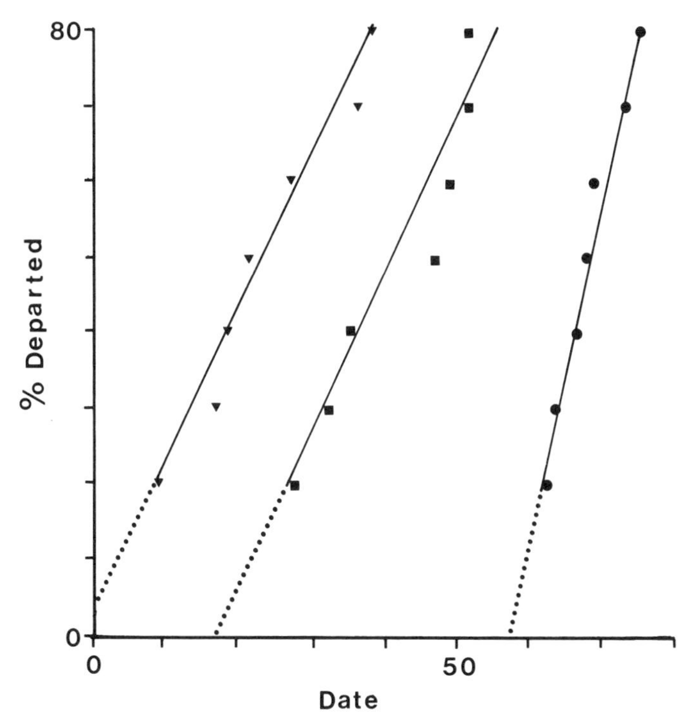

Figure 5.5 The relationship between the proportion of the Bewick's Swan flock that had departed from a wintering site in England at different dates under three different light regimes. Dates are the number of days after January 1. Light regimes: triangles—illumination until 22.00; open circles—illumination until 20.00; dots—natural daylength. From Rees (1982).

again to the wintering area, indicating that they had, because of the unnatural lighting, migrated at the wrong time.

The slope of the regressions of departure against daylength is significantly shallower in the early departures. The actual daylength at the time when the first swans left the area (the x intercepts on Figure 5.5) varied in each case, from 13.7 hours in the early group to 10.8 hours under natural conditions. Both of these effects may be due to the inhibitory effects of other environmental or physiological regulators.

5.4.2 *Weather conditions*

Once the daylength threshold has been reached, prevailing weather conditions determine exactly when the birds will depart. The danger of air strikes over Winnipeg airport in central Canada prompted a major study on the migration of Lesser Snow Geese in the early 1970s (Blokpoel and Gauthier 1975). A number of weather variables which might affect migration timing were examined, and the birds were found, not surprisingly, to migrate preferentially under conditions of following winds, though wind strength had no apparent additional effect. The birds also showed a preference for periods with lower than average precipitation.

5.4.3 *The extent of body reserves*

Birds migrating great distances might start the journey on following winds, but they cannot predict what conditions might be like some hundreds of kilometres away, and particularly unfavourable weather can have a serious impact on their survival or productivity. Ebbinge (1989) showed that Brent Geese migrating from the Baltic to Siberia bred poorly, despite being in good condition on departure, when wind conditions were unfavourable. These birds (which have a lean weight of around 1 kg) probably need about 250–300 g of fat to fuel the journey in still air; this would be increased by 50% with 30 km/h head winds. Considering that the total amount of fat available in body reserves is in the region of 600 g, differences in migration conditions can have a substantial impact.

One might expect that since body reserves are so important in spring (see also Chapter 2), the timing of migration might depend to some extent on body condition. There is no evidence that this is the case in Ebbinge's study. Brent Geese left the Netherlands at the same time each year, irrespective of the extent of their reserves. Presumably the most potent selective pressure is for the correct time of arrival, since any delay in nesting in the high Arctic will inevitably mean a sacrifice of that year's breeding effort (see Chapter 6). The extent of artificial feeding similarly did not have a significant effect on the departure dates of swans (Rees 1982). Recent unpublished evidence on Barnacle Geese, however, indicates that birds in poor condition depart later from the wintering grounds and from migration staging areas than those in good condition, as assessed by their abdominal profile (Owen 1981).

5.4.4 *Altitudes and speeds*

Sea duck migration over the sea is usually at low altitude; streams of birds can be seen passing headlands, almost skimming the water. When crossing land, however, Eiders in the Baltic sometimes fly at heights above visible limits (Alerstam *et al.* 1974). The median altitudes of geese migrating over land seems to be around 600 m above ground level for both Snow Geese (Blokpoel 1974) and Canada Geese (Bellrose 1978), though altitudes commonly varied between 300 and 1500 m. There are several exceptional records of high altitude flights; Bellrose records a pilot report of Snow Geese flying at 20 000 feet (6200 m) over Louisiana.

The most remarkable record comes from the detection of a radar echo at over 8000 m moving south over Northern Ireland in 1967 (Stewart 1978). A plane was diverted to investigate and the pilot verified the record as a flock of about 30 swans (presumably Whoopers migrating from Iceland). The altitude was 8200 m—the highest ever recorded for a bird. In 1988 however, a large jet collided with an object over the Atlantic off Newfoundland at a height of 35 000 feet (10 600 m) and had to make an emergency landing. There was a dent in the nose, but no abrasion, suggesting a soft object, possibly a goose.

Waterfowl have very high wing loading (body weight in relation to wing area) and have to fly at great speeds to avoid stalling. The Eider Duck has the highest wing loading and is thought consequently to have the highest stalling speed and to be the fastest flying bird (Rayner 1985); no accurate measurements of its flight has been made, but figures given in Alerstam *et al.* (1974) suggest air speeds of up to 90 km/h (56 mph). Geese generally fly at around 60 km/h (40 mph) in still air, but ground speeds of up to 185 km/h (115 mph) have been recorded for Snow Geese, and Whooper Swans were travelling at 140 km/h over the ground, assisted by a jet stream.

As we have already discussed, waterfowl choose to set off on migration under favourable wind conditions and their choice of altitude clearly affects the benefit they get, both from the decrease in 'drag' in the thinner air at high altitudes and variations in wind speed with height. It seems likely that the birds actively seek the most favourable conditions, which give the most energetic benefits. Their capacity to do this, and their extraordinary physiological capability is illustrated by the Whooper Swan example, further investigated by Elkins (1979). The temperature was in the region of −48°C and the jet stream was running at up to 100 knots (180 km/h). The low temperature would not be detrimental; it might even help since flight generates a considerable amount of heat which the birds have to dissipate. The feat of flying in the rarefied atmosphere at 8200 m is, however, very considerable. The benefits in energy saving are, of course, enormous; with a following wind of 180 km/h the birds could theoretically migrate in a quarter of the time it would take them in still air.

5.5 Dispersal

After the breeding season animal populations are at their largest and there is greatest potential for competition. Subordinate animals are displaced and post-breeding dispersal is a common phenomenon. We also deal here with the movement patterns from year to year and the likelihood of the same individual returning to the same breeding, staging or wintering area year after year.

5.5.1 *Age and sex differences*

The birds that do move out of their natal area are usually young birds; all of seven swans that moved outside the range of an isolated population of Mute Swans in Scotland were in their first year (Spray 1981).

Migratory species disperse more widely and the young of the year tend to move sooner and further than adults. For example, young Gadwall disperse southwards into France in their first winter, whilst others from the European mainland move into Britain to winter (Owen *et al.* 1986). The mortality rate of young is presumably less if they move than if they stayed within their natal range and competed with more experienced birds. Dispersal is also important for a species to discover new habitats and extend its range, as the Gadwall has done both in Europe and in North America (see page 94).

Geese and swans pair for life and the sexes migrate together and stay as a pair, but the migratory patterns of the two sexes of many ducks are radically different. In an analysis of five species of European dabbling ducks, the males, which take no part in incubation, were found to migrate southwards during early incubation, initially to a moulting area and subsequently to northerly parts of the winter range. The females lag behind and moult on the breeding grounds, but catch up with the males by early winter and then overfly them to winter in larger proportions in southerly parts of the winter range (Perdeck and Clason 1983). In the spring duck species differ, depending on the earliness of pairing and breeding but the sexes meet up again in late winter and spring. This 'leapfrog' migration pattern leads to a disparity in the sex ratio in at least parts of the range, and this is discussed in more detail in Chapter 6. As an example of the different movement patterns in the two sexes, the median recovery locations in different months of Wigeon ringed in Britain are shown in Figure 5.6.

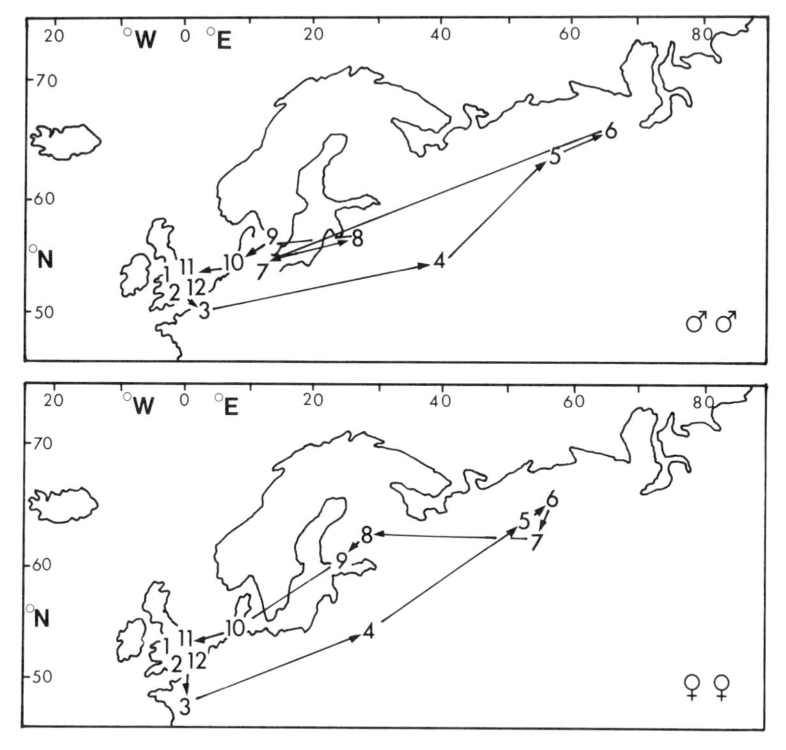

Figure 5.6 The median position of recoveries of British-ringed Wigeon, in different months (numbers), males above, females below. From Owen and Mitchell (1988).

5.6 Philopatry

5.6.1 *Breeding sites*

Where population intermingle during winter there is an opportunity for individuals from different breeding areas or sub-populations to pair together and cause mixing of those segments. In waterfowl it is the female that usually returns to her natal area to breed and the male follows. The phenomenon whereby males become displaced in this way is known as 'abmigration' and is a common feature in duck species. Female Buffleheads tend to return to their natal lake year after year and even use the same tree hole for nesting (Erskine 1961), and a high degree of homing to the natal area is probably also true of Long-tailed Ducks (Alison 1977). In the

Goldeneye, fidelity to the nest has been linked to higher reproductive success (Dow and Fredga 1983).

In geese and swans female loyalty to the natal breeding area is also the rule. It is well documented in the Lesser Snow Goose, where the return rate to the colony of origin is almost invariable for females and rare for males, which by chance are likely to pair with a partner of different origin (Cooke *et al.* 1975). It is not as clear cut in the Mute Swan, though there is a clear tendency for females to nest near their area of origin (Coleman and Minton 1979), as do those of the Canada Goose (Lessells 1985).

5.6.2 *Wintering and staging areas*

Individuals also tend to be traditional in their use of wintering places and staging areas on migration. Extensive marking of Dark-bellied Brent Geese in Europe showed that individuals tended to be loyal to the same wintering site, not only the same estuary but the same feeding area within an estuary, year after year (St. Joseph 1979). There are also distinct patterns of movement of different groups; Brent wintering in France stage in the Netherlands in spring, whereas the fattening areas of the birds wintering in England are on the German part of the Waddensee. Wintering Canada Geese can also be separated into subunits using different roosts and feeding areas and returning there from year to year (Raveling 1979). There is great loyalty to spring staging areas too, with individual Barnacle Geese returning to tiny parts of the staging areas, sometimes only a few square metres in extent, year after year (Gullestad *et al.* 1984). A similar pattern of site loyalty has been described for the Greenland White-fronted Goose (Wilson *et al.* in press). Birds ringed in a very restricted part of the breeding area have been seen throughout the whole winter range. However, individual geese are almost wholly traditional to their wintering site once they have matured and paired.

5.6.3 *Who leads and who follows?*

It is clear that it is the female that makes the choice of breeding area, but there is very little evidence as to which sex dominates in the choice of wintering site. In a detailed study of the Bewick's Swan Rees (1987) has produced the only evidence on this point. She determined the wintering site chosen by a newly established pair in relation to the partners' previous experience of wintering areas. Males were predominantly responsible for determining both the area chosen and the timing of the migration to the wintering grounds. She also analysed the pre-flight behaviour of pair members during their stay at a wintering area and found that in autumn it was the male which initiated movement, whereas in the spring there was a switch and the female began to lead. This falls neatly into the predicted pattern of female determination of the pre-breeding and location ecology and the male taking responsibility for autumn migration and wintering areas.

5.6.4 *The significance of site loyalty*

The patterns described in this section illustrate that waterfowl are extremely predictable in their movement. Why are their movements so rigid and traditional? As we have described in Chapter 2, waterfowl have specialised and well-defined habitat and food requirements. They also necessarily operate on a very strict annual cycle and have little room for error if they are to fatten, migrate and breed within the very short time they have available in the short Arctic summer. In species which use a number of different areas through the year, the accumulated experience of the local resource distribution and sources of danger are important for individual survival and breeding success.

There are social advantages in using traditional areas too; geese and swans pair for life and have an extended family life (Chapter 4). If members of a family become disunited during the winter they find each other at the traditional roost. The evidence for subunits of populations staying together throughout the annual cycle suggests that there are differences between species. In some populations, however, like the smaller groups of Canada Geese, traditional behaviour may lead to a high degree of genetic isolation, essential to maintain adaptive traits (Raveling 1979).

5.7 Seasonal home ranges

Within a season, waterfowl move around in a more or less predictable way within a home range, the characteristics of which vary with the species. This section deals with the mobility of waterfowl within a season.

5.7.1 *The breeding season*

Geese and some swans are generally territorial during the nesting season, but may move in search of food during the rearing period. Mute Swans remain on a defended territory throughout the pre-fledging period and the size of the stretch of river or lake they defend depends on the quality of the habitat; the birds defend only as much as is necessary to provide enough food for successful rearing of the brood (Birkhead and Perrins 1986). Geese move from nesting to brood rearing areas, sometimes a few kilometres distant, in the first few weeks of gosling life and thereafter move in a group among a series of ponds or stretch of shoreline until fledging, when they gather on more productive vegetation.

Egyptian Geese rearing broods are tied to water for safety and radiate from it in search of food. The regular incidence of broods along a lake shore suggested to Eltringham (1974) that families had individual home ranges which included a piece of shoreline to which they retreated when disturbed.

South American sheldgeese are highly territorial during the breeding season and defend territories like their ecological counterparts in the north. Male Shelducks do not guard the nest but defend a territory on a nearby wetland (usually estuarine mudflats) where the female feeds during her incubation breaks and where the brood is reared during the early days (Young 1970).

The movements of pre-breeding and breeding Pintails were monitored closely by Derrickson (1978). The ducks' mobility depended both on their status and on the stage during the nesting season. Paired males ranged over 900 ha, compared to 580 ha for unpaired males and 480 ha for females. The pattern of range size through the breeding period is shown in Figure 5.7. Twenty-nine per cent of Pintails showed no stability in their home ranges; the cumulative range size increased as they moved through the habitat (male in Figure 5.7). Of the others, 39% showed temporary stability, staying in one area for some weeks before moving to a new site. The cumulative range size of these birds shows a stepwise pattern (lower left figure in Figure 5.7). The remaining third of Pintails had rather stable home ranges which changed little through the prebreeding and breeding period (top left figure). Although some of the variability in the mobility of Pintails is explained by sex and time, there is a large amount of individual variation which is probably due to local variations in the food supply and perhaps competitive interactions with other ducks.

Pintail are among the more mobile of ducks throughout the year and at no time do individuals defend a food resource. There are many other

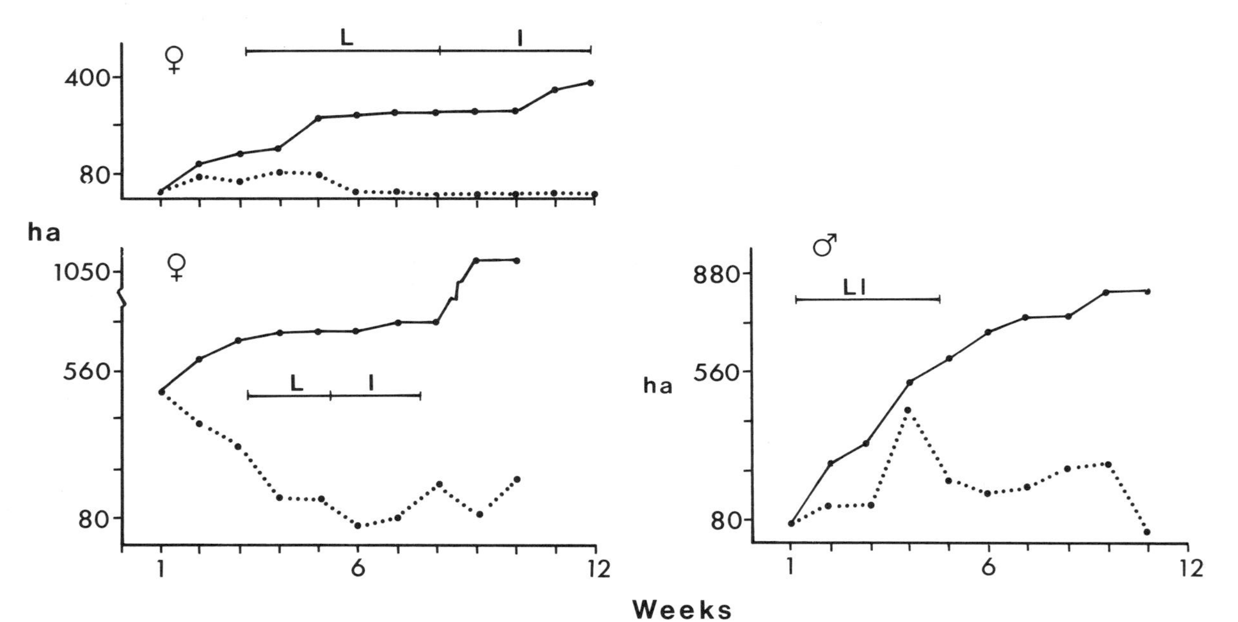

Figure 5.7 The relationship between the timing of breeding and home range size (in hectares) of three individual Pintails in North Dakota. The upper line in each graph is the cumulative home range i.e. the total area over which the bird wanders and the lower line indicates the extent of the range in each 7-day period. Laying (L) and incubation (I) periods are indicated on horizontal lines. From Derrickson (1978).

strategies among the ducks which range from highly territorial to highly mobile and these are discussed in more detail in Chapter 4.

5.7.2 *The winter season*

Winter home ranges also vary considerably between species. Nearly all waterfowl have at least two components to their winter habitat—a roost and a feeding area—and they generally make regular flights between them. There are also in some cases subsidiary resting areas where the birds go to rest and preen. Working on Teal in France and in Louisiana, Tamisier (1974, 1985) developed the concept of the 'functional unit', where ducks moved within a unit which consisted of a roost and several feeding areas. He had evidence from studies involving telemetry that individual Teal predominantly stayed in one functional unit and that these are in some senses social units which overlap little. One of the benefits of such a system is that the communal roost acts as an 'information centre', whereby individual animals can benefit from the experience of a group and can more effectively exploit the food resources (see page 84). Other work on ducks has not supported the concept of the functional unit; most studies of Teal have represented the species as being highly mobile (Ogilvie 1981) or having rather limited and short-term site tenacity (Baldessarre *et al.* 1988). These studies have, however, been much less intensive and on a wider geographical scale than those of Tamisier, whose ideas on the social nature of home ranges in wintering ducks have not yet been rigorously tested.

The species which tend to be highly site loyal from year to year also tend to be conservative in their home range within a season. An extreme example of site loyalty shown by Greenland White-fronted Geese is described by Wilson *et al.* (1990). The home ranges of two individual neck-banded birds over several years on the Wexford Slobs in Ireland are shown in Figure 5.8. Clearly the birds are extremely faithful to individual feeding places from day to day; there is evidence that they also have localised roosts, so that the system is similar to the 'functional unit' proposed by Tamisier (1974).

In some populations, as already mentioned, there are links between aggregations at different times in the life cycle, but it is uncertain whether even these apparently stable groups operate as a social unit. In some species, at least, it appears that each individual or family acts independently in its movement and coincidentally or for geographical reasons associates with others. This aspect deserves more study.

Figure 5.8 The positions of sightings of two individual Greenland White-fronted Geese at the Wexford Slobs, Ireland in several winters. Lines within the area mark the boundaries of the main water bodies. The area is, in addition, split into individual fields by ditches, which are not shown. From Wilson *et al.* 1990.

5.8 Unpredictable movements

In some circumstances animals move in response to local and unpredictable changes in food availability. The most common shifts as far as waterfowl are concerned occur in response to drought or hard weather. Movements in response to drought concern both breeding and wintering. Spring droughts in the prairie breeding grounds of ducks in Canada can cause massive northward movements in search of suitable wetlands (Hansen and McKnight 1964). There is an inverse relationship between the number of dabbling ducks in the prairies and in the north; in severe drought years up to 6 million dabbling ducks are displaced. The movement is undoubtedly vital for survival, though breeding success of the displaced birds is low. In Pintails, there was an inverse correlation ($r = -0.92$) between the proportion of birds displaced and an index of annual breeding success (Smith 1970).

Many species which winter in subtropical areas where habitats are unpredictable show this pattern of movement regularly in response to local conditions. The movements of the Australian Grey Teal have been described earlier in this chapter; many species wintering in Africa and arid areas of the Americas and Asia show similar movement patterns.

The most well-studied erratic movements are those brought about by cold winters in temperate areas. Massive southerly movements occur, and inland waters are deserted as waterfowl move towards the coast. The severe winter of 1962–63 was the worst in Europe in this century and ducks from areas bordering the North Sea moved southwards to France and Spain and westwards to Britain and Ireland. Most geese deserted the Netherlands for France (Philippona 1966) after a week or two of cold weather. The immediacy with which movements occur depends mainly on the body size of the species and the wintering habitat. Geese and swans have sufficient body reserves to withstand many days without food, but the smaller ducks must leave almost immediately. Its small size and requirement for shallow water for feeding makes the Teal the most vulnerable species (Ogilvie 1981). Females are smaller than males and suffer more in severe cold and probably also in competition with males for limited food. In hard winters in Texas the ratio of male to female Green-winged Teal was 2.8:1 whereas in the population as a whole it is around 1.3:1. The difference was due to rapid movement of females out of the area rather than to mortality, since the sex ratio of dead birds was similar to that in the live population (Bennett and Bolen 1978).

5.9 Implications for conservation and management

Migration is a necessary aspect of the life history of waterfowl, and because of the patchy nature of the habitat, journeys are highly traditional in their timing and routes. Losses of regular staging areas can be disastrous, especially for populations relying on small, isolated habitats to break long journeys. International legislation is needed to safeguard flyways and this is now being put into place (see Chapter 7).

Species which have small home ranges are unable to respond as quickly to changing circumstances as are more mobile nomads. The conservative Greenland White-front disappeared from many of its former haunts when its habitat disappeared and the population overall declined in the 1960s. Subsequent increases in the population occurred very largely by increases in the traditional major haunts rather than by expansion of range and the pioneering of new areas (Wilson *et al.* 1990). Home range behaviour also has implications for management in relation to agricultural damage. Mobile species are more easily moved on by scaring than more sedentary ones, which shift rather short distances to nearby fields or farms. Clearly the movement patterns of different species must be understood if they are to be sensitively managed.

Waterfowl are vulnerable in cold weather and bans on hunting are regularly enforced in many countries, based on some threshold level of severity (Batten and Swift 1981). If populations are to be safeguarded however, the birds must be protected in the refuge areas to which they retreat even though these may not be subject to severe weather. Recent studies in Europe have aimed to identify such areas so that they can be given protection under international legislation (Ridgill and Fox 1989).

CHAPTER SIX
POPULATION DYNAMICS

6.1 The nature of populations

A population is a group of animals inhabiting an area which is to some extent discrete from other groups of the same species. The number of animals in a population can change by the recruitment of young, mortality, immigration and emigration. Some populations are closed, having no interchange with any others and these change only through mortality and recruitment. Examples would be the sedentary inhabitants of isolated islands, many of which have evolved from the ancestral stock so that they are distinct enough for subspecific status. Other groups may be so isolated geographically or by tradition that they rarely come into contact with others. Many geese fall into this category.

Many species of migratory waterfowl consist of several groups which can be termed populations, but which do show a degree of genetic mixing. Most commonly, there are several somewhat discrete breeding areas whose populations mix in winter and, through pairing on the wintering grounds, genetic mixing takes place. These breeding groups usually travel over discrete flyways and can be termed sub-populations and form a convenient management unit (see Chapter 7).

The understanding of population processes provides the key both to the exploitation and conservation of animals; changes in prevailing conditions may cause numbers to decline or increase. Knowledge of the 'bottlenecks' which are limiting populations can also be useful if by management they can be removed and the population allowed to increase before stabilising at a higher level. Natural populations are limited by the availability of a specific resource, usually food. Numbers increase through breeding to levels in excess of the capacity of the resource and fall back again through mortality as the resource becomes limiting. Such mortality is said to be density dependent in that it is heaviest when numbers are initially highest. Density-dependent effects tend to damp down fluctuations in populations when a resource is limiting. Density dependence also operates on recruitment; competition for resources can limit the number of breeding birds, their hatching success or the survival of young.

6.2 Monitoring techniques

In some populations with limited ranges it is possible to assess the number of birds by direct counts but, more commonly, estimates must be based on sample surveys and extrapolation. In Europe and North America an attempt is made to assess the size of the post-breeding waterfowl populations on an annual basis, and this is being extended to other parts of the Northern Hemisphere. Estimates of the late winter populations are much less often made and the losses over winter difficult to estimate directly.

In North America aerial surveys on the breeding grounds aim to assess the number of breeding pairs, but the vital information on the proportion of birds breeding, nest and fledging success and the factors which control them require detailed studies on segments of the population.

Estimates of the proportion of young in flocks of geese and swans can be made in the field in the autumn since, in nearly all species, it is possible to distinguish birds of the year from adults at a distance. These age ratios, coupled with counts of the population each year, can be used to estimate the annual mortality rate (see e.g. Boyd and Ogilvie 1969). However, these estimates are highly sensitive to error and should be used with caution (Owen 1980a, pp. 151–155). Mortality estimates can also be made from ringing recoveries based on the pattern of recoveries over succeeding years. There are problems with these methods too, mainly to do with variations in both the recovery and the reporting rate, i.e. the likelihood that birds will be found and reported. Because a high proportion of the deaths of waterfowl are the result of hunting (see Stoudt and Cornwell 1976), the recovery rates are relatively high—up to 25%. The problem with this is that the chances of recovery are linked with the cause of death—natural mortality is scarcely reported, especially in remote Arctic regions.

More recently, at least in some populations, more direct methods of studying population dynamics have come into widespread use. These depend on marking birds with large rings or neck-collars which are readable from a distance with a telescope and enable multiple records of live birds to be made and detailed histories of individuals' productivity and mortality built up. These are most effective when used on rather small populations over a limited geographical range, but they do provide clues as to how other populations behave.

The following account presents data from a wide variety of studies and using a large number of different techniques to try and compile a picture of how numbers are controlled at various stages of the life cycle and what factors bring about changes in population size.

6.3 Recruitment

The stage at which an animal can be said to contribute recruits to a population is when its young begin to breed themselves. Since very few studies are long enough or detailed enough to determine this, the production to some intermediate stage must be used as a guide. An important rule to remember is that for a population to remain stable each bird that reaches breeding age must, on average in its lifetime recruit one individual to be part of the breeding stock.

6.3.1 *Age at first breeding*

The generation time is an important aspect of population productivity; clearly a species that breeds first at 1 year old has a much greater productive capacity than one that delays maturity until the age of 2 or 3. In general a number of reproductive traits are correlated. The age at maturity and the survival rate of young to fledging increases with the size of the bird whereas clutch size generally decreases (see page 48).

The age of first breeding is affected by physiological and ecological or behavioural factors. Although a species may be physiologically capable of breeding at a certain age, it may not do so because it has to gain experience in the process of mating, nesting or in gaining the correct diet and sufficient of it for successful reproduction. Not only does age affect whether a species or an individual nests or not, but it also affects how successful it is at various stages—the implications of age will be considered at each stage in the following account (see also Chapter 3). Without the ecological and density constraints, the performance of a species in captivity gives an indication of the earliest age at which a species can breed. Information from captivity and from the wild is combined in Table 6.1 to give an indication of the age of physiological maturity and the age at which breeding normally occurs in the wild for a number of representative waterfowl species, in relation to their body size.

Table 6.1 The earliest age at which a number of representative waterfowl breed and the range of ages at which they nest in the wild, in relation to their body size.

Species	Female body weight (g)	Minimum age (years)	Normal in wild (years)
Mute Swan	9,000	2	3–5
Snow Goose	2,500	2	2–4
Barnacle Goose	1,700	2	2–5
Upland Goose	2,800	2	?
Common Shelduck	950	2	2–4
Wood Duck	700	1	1–2
Mallard	1,100	1	1
Blue-winged Teal	380	1	1
Canvasback	1,150	1	1–2
Tufted Duck	700	1	1–2
Common Eider	2,000	2	2–4
Velvet Scoter	1,200	2	2–3
Goldeneye	750	2	2–4
Goosander	1,200	2	?
Ruddy Duck	600	?2	?

6.3.2 *Non-breeding*

In most populations some birds which are old enough to breed and are paired do not nest in some years, but the extent of non-breeding is very difficult to determine. In Mute Swans in England about a third of paired birds on territories did not breed, and this was a rather constant proportion from year to year (Minton 1968). In addition non-territorial birds were also not breeding. There could be a number of reasons for this, but most likely is the inability of some birds to reach the necessary body condition for breeding. Reynolds (1972) demonstrated that Mute Swans of both sexes had to exceed a threshold weight (10.6 kg for males and 8.8 kg for females) before they were able to nest. Black Swans in Australia gather in very large concentrations and nest in colonies on the vegetated margins of lakes. A high proportion of non-breeders, including mature swans, are usually present; in one year only 400 nests were initiated by 2000–3000 swans. Shortage of nest sites is thought to be the main reason for non-breeding in this species (Frith 1982).

Considerable proportions of mature birds in migratory swan populations also fail to nest. In Alaska the number of pairs of Whistling Swans with broods in August varied between 15 and 48% (mean 31%) in different years (Lensink 1973). Since few pairs lose entire nests most of the lack of productivity is due to non-breeding.

Most goose populations that have been studied intensively also show variable degrees of non-breeding in different years. Non-breeding in Canada Geese is usually attributed to age or inexperience; Craighead and Stockstad (1964) assumed that two-thirds of two-year-old geese did not nest but that all those older than 3 years laid eggs. In Barnacle Geese, however, in the relatively early season of 1986, only 40% of mature pairs on a Svalbard breeding area nested (Owen and Black 1990). A number of factors can affect the proportion of non-breeders in goose flocks; among the most important are weather at time of nesting and density-dependent factors mentioned later.

The proportion of non-breeders in duck populations is more difficult to estimate, since generally only the breeding areas are surveyed. Krapu and Doty (1979) found that the vast majority of female Mallard of 2 years old or older, collected in the breeding season, had laid or initiated follicular development but that only a small proportion of yearlings were going to breed (they contained developing eggs). Exceptional weather conditions such as drought can cause a high proportion of ducks not to breed. In a North American prairie area, two-thirds of Lesser Scaups left the breeding area early and less than 5% nested (Rogers 1964). In those duck populations with a high preponderance of males there is inevitably a proportion of birds which cannot contribute to the breeding effort.

In ducks that nest in tree cavities, the fact that nesting density can be increased substantially by the provision of nesting boxes (see page 162) suggests that there is competition for nesting sites in many species, and this leads to high proportions of non-breeders. The fact that there is often a high level of communal or dump nesting (see e.g. Heusmann 1972) adds support to this (see page 72).

6.3.3 *Nest success*

Predation of eggs is a very important factor affecting the productivity of waterfowl (see Chapter 3). Different strategies are adopted by different species to minimise the effects:

a) nesting in inaccessible sites (islands, cliffs, tree holes)
b) nesting in cover and the female remaining on the nest when approached
c) spacing nests over wide areas to make detection more difficult and
d) nesting in colonies, where predators may be deterred by aggression by large numbers of birds or are 'swamped' by overabundant food.

There are few data on the success of the same species adopting different strategies, but predation rates in parts of the North American prairies, where duck habitat is so reduced that normally dispersed breeders have to nest in close proximity, is extremely high. Predators converge on dense concentrations and find most of the nests. Predation rates in geese are generally low; in Arctic Canada, losses of Canada Goose nests were of the order of 10% (MacInnes and Misra 1972). There are suggestions, however, that in certain years when lemmings, the main food of foxes in Siberia, are in short supply, the predators switch to Brent Goose eggs and cause complete breeding failure. The variation in predation rate has been proposed as a major determinant of cyclic breeding in the Brent (Summers and Underhill 1987). The evidence is, as yet, circumstantial, but there are correlations between Brent success and that of a number of wader species, which is suggestive. Fox predation was responsible for heavy nest losses in the Cackling Canada Goose in Alaska (D.G. Raveling pers. comm.). When numbers were high, predation had little impact, but following a reduction in the population and in colony size, similar numbers of predators had a greater relative effect.

Because waterfowl generally nest close to water, they are vulnerable to flooding. Goose nests in the Yukon Delta area of Alaska can sometimes almost be wiped out by high sea levels; the Brant, which nest in the lower areas next to the sea are subject to the heaviest losses, up to 90% in some years (Mickleson 1975). Duck and swan nests are more frequently affected by rising water levels; flooding is the second most important source of nest loss in Mute Swans in England (Birkhead and Perrins 1986).

The relationship between body weight on arrival and laying in Snow Geese is shown in relation to their clutch size and success in nesting in Figure 6.1. Females which had failed to hatch deserted their eggs when their weight fell below a critical level. Some females let their weight fall too far and died on the nest in late incubation. Nest losses through desertion can have a substantial effect on productivity. This relates to the extent of females' reserves, and the amount of food that is readily available near the nest. For example, the nest success rate of Barnacle Geese in 1986 varied from 21% on an island with little or no food to 84% in vegetated islands (Owen and Black 1990).

The hatching success of eggs which have been successfully incubated is normally extremely high. Amat (1987a) summarised data on infertility in eggs in successful nests from a large number of breeding studies of four species of dabbling ducks. The infertility rate ranged from an average 3.3% in Pintail to only 1.1% in Shoveler. Egg dumping normally leads to poor

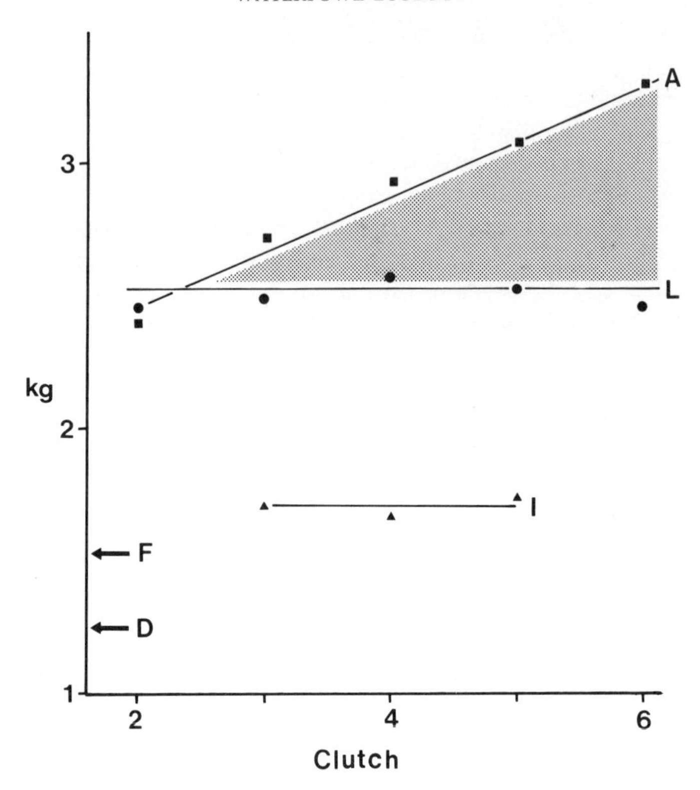

Figure 6.1 The relationship between the body weight of Lesser Snow Geese on arrival at the breeding grounds (A) and clutch size; the weight at the end of laying (L) and in late incubation (I). The weights of geese which failed (F) and those which had died of starvation on the nest (D) are indicated by arrows. The shaded area indicates the reserves on arrival, in addition to those needed for incubation, that females can devote to the clutch. Reproduced from Drent and Daan (1980); based on the data of Ankney and MacInnes (1978).

hatching success, from failure of fertile eggs to hatch rather than infertility. In Shelducks breeding in Scotland parasitised nests were frequently deserted. Even in those that were incubated to hatching success was low since some eggs were incompletely covered by the female or were exposed to suboptimal incubation temperatures (Pienkowski and Evans 1982).

In ducks and a few geese it is common for birds which have lost their nests early in incubation to re-lay and hatch the second clutch and rear young successfully. The contribution of renesting to population productivity can be substantial, especially when unusual conditions, such as a flood, cause widespread failure of the first nesting attempt. For example, 50% of Mallard nesting in Manitoba which lost their original clutches renested, and 40% of renesters reared broods (Dzubin and Gollop 1972). Renesting is considered in detail in Chapter 3.

6.3.4 *Survival from hatching to fledging*

Brood rearing areas for geese and swans may be some distance from the nest; the internal food reserves in the yolk sac are crucial. The young of Mute Swans can survive for up to ten days without food. Ducks also have large internal hatching reserves and these are essential to sustain them in the first few days whilst they learn to feed effectively (Kear 1965).

Swans are able to defend their young against almost all predators and predation is not an important factor in the survival of young Mute Swans (Birkhead *et al.* 1983). Geese are largely terrestrial feeders and are vulnerable to land as well as avian predators. The early stages of gosling life are the most vulnerable as the broods travel from the nest to the rearing areas. Cliff-nesting Barnacle Geese in Greenland are subject to 25% gosling losses to gulls and foxes as they leave the cliff ledges (Cabot *et al.* 1984). Young goslings are also vulnerable to chilling, and cold and wet weather in the first few days of life can give rise to considerable mortality. Losses after the first week are slight in most species, and these are largely due to predation. Three of 28 Barnacle goslings were taken by a fox in four weeks; predation occurred as broods moved between lakes (Prop *et al.* 1984). In areas with lemming populations, such as Wrangel Island in eastern Siberia, fox predation on goslings can be considerable in good lemming years when fox numbers are high. Losses between hatching and fledging can be as high as 40% (Bousfield and Syroechkovskity 1985). This is interesting in view of the supposed association of Brent losses (presumably nest losses) with poor lemming years.

Even though most ducklings stay on water between hatching and fledging, they are still vulnerable to predators. In parts of North America predation by pike accounted for 1.7 duckling losses per acre, or 10% of the seasonal production (Solman 1945). Mallard and Wood Ducks in Minnesota reared only 44% and 41% respectively of the young they hatched. Most losses were in the first two weeks, but their causes were not recorded (Ball *et al.* 1975).

Eider Duck broods suffer very high levels of predation by gulls; 80% of ducklings were taken by Herring Gulls in the first two weeks of life in the Ythan Estuary in Scotland (Mendenhall 1976). Ducklings were most vulnerable in bad weather when they had to feed more intensively, though bad weather in itself did not cause losses.

Many of the losses of young ducks are related to food, though much of the evidence for this is circumstantial. The home ranges of Goldeneyes are increased in size in areas where fish, which compete with ducklings for food, are abundant; hence the carrying capacity of the habitat is reduced. The use by broods of an experimental lake increased after the removal of fish (Eriksson 1979). The survival of Mallard ducklings is lower on newly established gravel pits than on natural lakes, and most of the mortality occurs during the first twelve days of life, when the ducklings are dependent on emerging chironomid fly larvae for food (Hill *et al.* 1987). They suggested that the difference was due to lower productivity of gravel pit habitats, and that competition from introduced fish exacerbated the situation.

6.3.5 *Post-fledging survival*

Birds are not truly 'recruited' until they enter the breeding population, and losses on the journey to the wintering grounds may be significant. These are, in most circumstances, difficult to quantify. There is, however, a negative relationship between the number of Snow Geese fledged on Wrangel Island and the proportion surviving to return the following summer, suggesting a density-dependent effect (Bousfield and Syroechkovskity 1985). Losses of young Barnacle Geese on migration were as high as 35% in 1986, when the density of families on the rearing areas was high (Owen and Black 1989a). Early post-fledging mortality is very difficult to investigate, but its importance has probably been underestimated, especially in species which undertake long, non-stop journeys over inhospitable terrain.

In populations which are limited by the extent of their winter habitat, which is the case in many ducks, there is a density-related mortality in winter, and the evidence suggests that, whether most of the mortality is a direct result of Man's activities or not, it is birds of the year that suffer most severely. This mortality occurs before the birds are recruited to the breeding population, but it is discussed in the section dealing with adult mortality (section 6.4.5).

6.3.6 *The factors determining recruitment rate*

The preceding sections have described some of the mechanisms affecting the productivity of waterfowl populations, but how do these interact with each other to control the number of young birds that a population produces? We examine some examples of studies which have looked at overall recruitment and their causes.

Weather has been implicated in the control of reproduction in many species, especially of those breeding in northerly areas. The amount of body reserves laid down by female geese in preparation for migration and breeding have long been known to affect breeding potential; the evidence is very clear in Snow Geese on an individual level (Ankney and MacInnes 1978). Their evidence for the relationship between body weight on arrival, clutch size and nest success is presented in Figure 6.1.

The most important determinant of fattening rate, especially in spring, is the weather as it affects the growth of vegetation in the wintering or early staging area and this can determine the productivity of a whole population. There is a highly suggestive correlation between the proportion of young Whooper Swans in Sweden in autumn and the severity of the preceding winter, as shown in Figure 6.2. Winter conditions account for 58% of the variation in breeding success. Conditions in May on the breeding grounds are also important, accounting for 32% of the variation in productivity. The relationship in winter could be brought about by a correlation between weather then and conditions at other times, but the relationship between winter temperatures and those in the following May was weak and nonsignificant. It is suggested that the effect on productivity is brought about by the effect of cold weather on food availability and thus on the condition of females (Nilsson 1979).

The body condition of Brent Geese when they departed from their last staging areas in north-west Europe was a good predictor of the breeding success in most seasons (Ebbinge 1989). Wind conditions on the journey north caused a depletion of body reserves in some springs, depressing success. Prairie droughts are also implicated in poor breeding seasons of Snow Geese, again by limiting reserves which result in lower clutch size and hatching success (Davies and Cooke 1983). Such relationships have not been described for ducks, whose habitat may not be available at all in drought years, but they must apply at least to some extent to those species, such as the Eider Duck, which rely heavily on their internal food reserves for egg formation and incubation (see Milne 1976).

Weather on the breeding areas is an important variable in the breeding success of all waterfowl, since water is a crucial element of the habitat. As has been mentioned above, droughts can inhibit breeding altogether, whilst spring temperatures and snow cover cause a delay in nesting and cause a reduction in breeding success. In all species which have been studied in

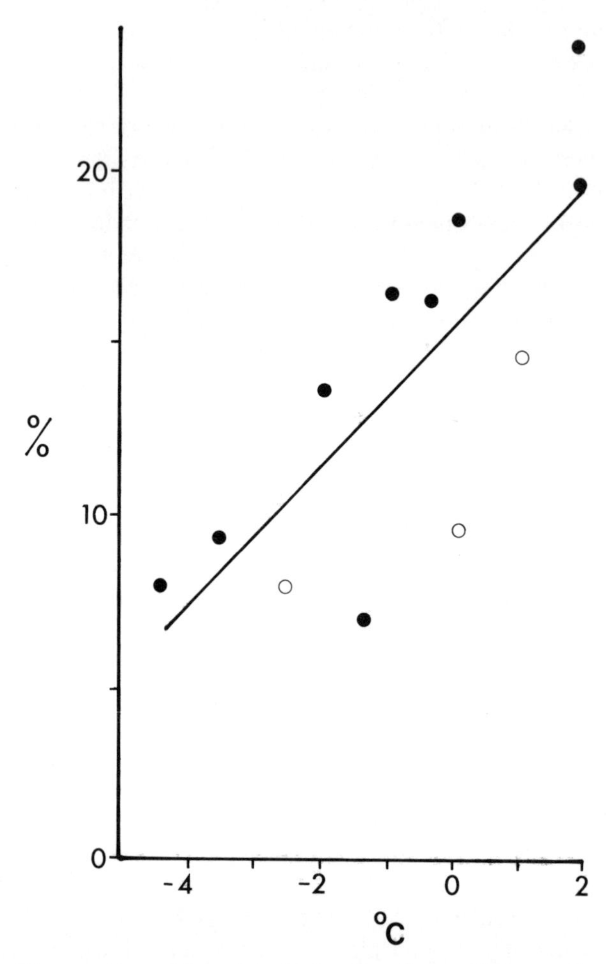

Figure 6.2 The percentage of first-winter birds among flocks of Whooper Swans in south Sweden in January in relation to mean temperature for the preceding December–March at ten meteorological stations in south Sweden. Correlation coefficient $r = 0.76$, $P < 0.01$. Open circles are ones that were followed by a late spring in the breeding area. From Nilsson (1979).

detail, clutch size decreases as the season progresses. This is partly because depletion of body reserves necessitates a reduction (see page 118) and partly to avoid delaying hatching unduly (late-hatched young generally survive less well).

Failure to incubate successfully increases in geese and swans in late seasons. Since reserves are used for maintenance energy prior to nesting, females leave the nest to feed more often, finally to the extent of desertion and/or predation of the unprotected eggs. Nesting success (at least one young hatched) of Barnacle Geese in Spitsbergen was 15–25% in late years compared with 60–80% in early ones (Prop *et al.* 1984). This relationship is probably less strong for ducks, except those, such as the Eider, that are goose-like in their laying and incubation habits. Weather effects on the food supply of ducks can have considerable impact on breeding success in many species, especially those that depend on matching the hatching date with the emergence times of insects; the survival rate of young ducklings in Myvatn, Iceland, is closely related to the abundance of chironomid flies (Bengtson 1972).

The effect of seasonal differences on the breeding success of small Canada Geese in a Hudson Bay nesting area is shown in Figure 6.3. Lower clutch size and high early losses of young contribute most to the difference in reproductive performance; the number of young fledged per female in a late year is less than half that in an early one.

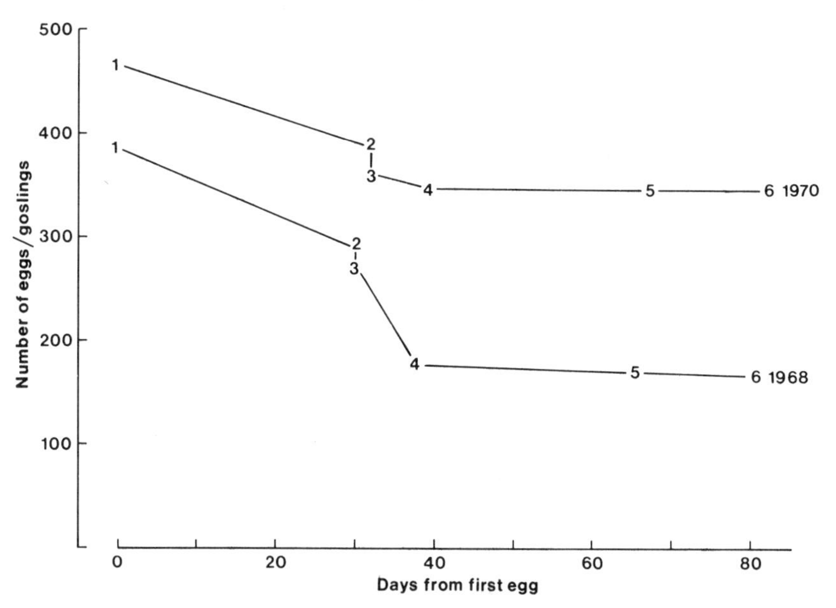

Figure 6.3 The reproductive performance of small Canada Geese in early (1970) and late (1968) seasons at the McConnell River, Northwest Territories, Canada. Numbers: 1—eggs at laying; 2—eggs at hatching; 3—goslings at hatching; 4—young at about 1 week; 5—young at about 5 weeks; 6—young at fledging. From figures in MacInnes *et al.* (1974), reproduced from Owen (1980a).

The age structure of a population also affects its breeding potential, since young and inexperienced birds tend to be less successful than more mature ones and in the longer-lived species do not attempt to breed until their second or third year. The effects of age and experience on breeding success are dealt with in detail in Chapter 3.

Competition for nest sites and food increases with density of breeding birds, but there is little evidence of density-dependent effects on breeding success in waterfowl, except perhaps in hole-nesting species where there is a scarcity of suitable nest sites (see page 163). It may be that waterfowl populations have been regulated in the past by food resources on the wintering grounds, such that numbers in spring were reduced by winter starvation to a level well below the capacity of the breeding areas. We argue (Chapter 2) that this was the case before Man interfered with wetlands.

Because the limit on wintering habitat for most goose populations has been removed as the birds have taken to feeding on the practically unlimited food resources on agricultural land, density on some nesting areas has increased and there are indications of density-dependent effects on recruitment. In Lesser Snow Geese nesting in La Perouse Bay, Manitoba, clutch size has decreased progressively as numbers have increased and the geese have started to have an impact on their food resources (Figure 6.4). The body size of young geese also decreased with the increase in the colony, and the juvenile survival rate declined (Cooke 1990). Both these and the decline in clutch size suggest that density-dependent effects are influential in spring staging areas or on the breeding grounds. The population of Barnacle Geese breeding in Svalbard is also showing density-related declines in the proportion of birds breeding, nest success, and in juvenile and adult survival (Owen and Black 1990).

To summarise, weather during the pre-breeding and breeding period is the most important factor causing variability in recruitment of young waterfowl from year to year. Which individuals contribute to the year's production depends on a number of factors, the most important of which is probably age, though in territorial species dominance (the ability to defend the resource) may be equally important, though this is probably related to age. In most migratory populations numbers are limited by the supply of winter food, density-dependent effects on recruitment are seldom demonstrated. There are, however, few studies on the highly dispersed and territorial species such as the migratory swans and sheldgeese, whose populations may be more commonly controlled in a density-dependent fashion.

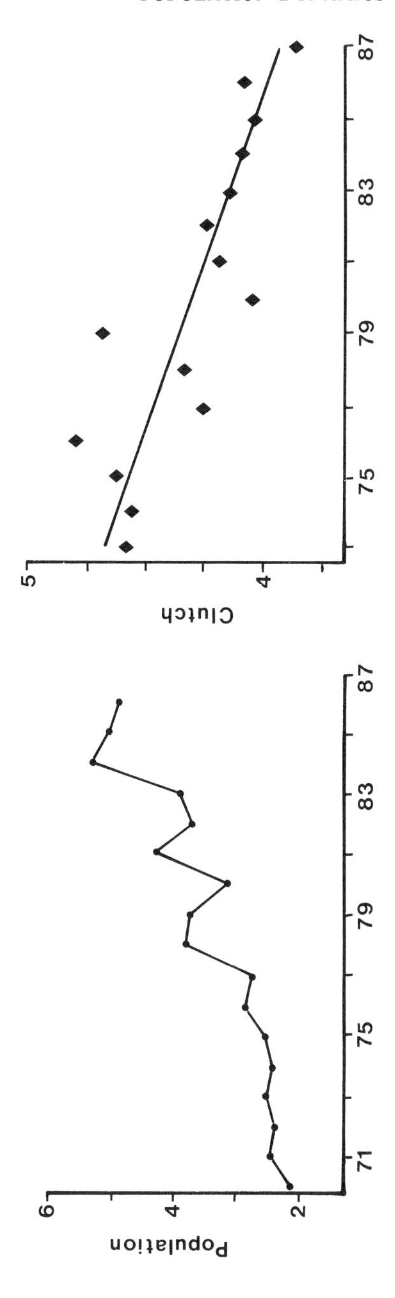

Figure 6.4 The estimates of the size of a Lesser Snow Goose colony at La Perouse Bay, Manitoba (left) and the mean clutch size (adjusted for nest parasitism and laying date), for the same period (right). From Cooch *et al.* (1989).

6.4 Mortality

There are only three major natural causes of mortality in animals—disease, starvation and predation. We do not assume here that the three factors act entirely independently. For example, an animal which is weakened from lack of food is more susceptible to disease, which may be recorded as the cause of death. For example, periodic die-offs of Greater Magellan (Upland) Geese in the Falkland Islands were put down to a combination of food shortage and parasitic infections which were sometimes the direct cause of death (Harradine 1982). Diseased birds may also be more vulnerable to predation. Predation risk can also be linked to the food supply; when food is in short supply an animal may increase its risk of predation, by venturing into more dangerous areas to find food or by decreasing its vigilance; it may be misleading to consider predation and starvation in isolation from each other.

Accidental death (e.g. collisions with natural objects and the direct effects of weather) may account for small numbers of waterfowl. Natural causes of mortality may also interact with Man's activities. For example, artificially large concentrations of birds on roosts may contribute to disease risk and certainly aid its spread. Having made these provisos we will examine each of the main causes of natural mortality in waterfowl before discussing how mortality interacts with recruitment and population size to control numbers. Waterfowl mortality is affected to a large extent by the influence of Man, and these effects are treated separately.

6.4.1 *Disease*

As we pointed out at the beginning of this chapter, it is very difficult in waterfowl to determine the importance of natural causes of mortality since the recovery rate is very much lower than that for birds dying from shooting or other Man-induced causes. Corpses are soon cleared up by predators and birds dying naturally in small numbers are rarely picked up.

Stoudt and Cornwell (1976) undertook a massive survey of the non-hunting mortality of North American waterfowl. They used data on the causes of death from those reporting ringed birds, and questionnaire surveys, as well as accumulated data in refuge and other reports and files. Some of their nonhunting mortality classifications were Man-related and we have separated these from the truly natural causes in the following account. Of the 1.8 million deaths of natural causes, 99.3% had died of disease including major outbreaks of botulism (92.6% of all deaths) or fowl cholera (5.5%), when large numbers of birds were involved and the incident was obvious and well recorded (this tends to overestimate the importance of disease).

Botulism is caused by a lethal neurotoxin produced by the bacterium *Clostridium botulinum*, an inhabitant of warm mud. The disease also affects Man. Outbreaks in waterfowl have been recorded from most of the world (review in Smith 1976). The disease organism flourishes where shallow water areas dry out in hot weather; 90% of the disease losses recorded by Stoudt and Cornwell (1976) occurred in summer and autumn. The largest outbreaks have involved more than 100 000 deaths, and the problems are exacerbated by the great concentration of birds on refuges. Most of the outbreaks are on federal refuges; dabbling ducks, because of their feeding in shallow water, are most severely affected.

Fowl cholera, which is caused by the bacterium *Pasteurella multocida*, is a disease of wintering waterfowl. The natural extent of the disease is not known, but 'overcrowding' of birds because of artificial factors has been responsible for most of the outbreaks, some of which have been estimated to kill up to 60 000 birds (Vaught *et al.* 1967).

Duck plague or Duck Virus Enteritis (DVE) is a disease whose importance has been recognised only recently. The virus is known mainly from domestic and captive waterfowl, but it can cause significant deaths among wild birds. The first record of the disease in the wild was in 1967, but no major outbreaks occurred until 1973, when 40 000 ducks, mainly Mallard, succumbed in South Dakota (Friend and Pearson 1973).

6.4.2 *Starvation*

Starvation as a result of food shortage is the main factor limiting the size of naturally regulated animal populations (Lack 1954). Evidence of starvation is, however, not usually obtained; birds tend to simply 'disappear' when times are hard. Many quarry populations are heavily harvested and some are kept below the carrying capacity of the habitat, so starvation would not be expected to affect them. However, in those populations which are density-regulated, death from starvation should be commonplace.

Starvation was not specifically identified as a source of mortality in the Stoudt and Cornwell (1976) survey. Cold weather was, however, said to be responsible for 0.6% of the deaths. It is during periods of cold weather that deaths from starvation are evident. The winter of 1962–63 was the coldest (as defined by the total of negative daily mean temperatures) in Europe in this century (Ridgill and Fox 1989). Even the final wintering areas of many species, such as Britain and France, suffered severe coditions. Large numbers of birds were picked up having died of starvation in Britain (Boyd 1964b). The overall impact of these losses is difficult to quantify, but a drop in the winter index of Mute Swans indicated that several thousand had succumbed (Owen *et al.* 1986). More than a hundred swans were found dead at one site (Perrins and Ogilvie 1981). In the cold winter of 1976–77 many Atlantic Brant died of starvation and the population was halved as a result (Rogers 1979).

Cold winter weather is responsible for reductions in the populations of waterfowl in Sweden; numbers in years following hard winters are severely reduced. The magnitude of the effect depends on the species; those wintering in shallow water fare worst. Following the particularly severe winter of 1978–79 the population of Coot suffered an 80% decline over numbers in the previous autumn (Nilsson 1984).

Food shortage causes mortality which may not, at first sight, be perceived to be due to starvation. For example, the mortality of Barnacle Geese on migration, described by Owen and Black (1989a), is the result of failure of some birds to lay down sufficient body reserves to fuel the long direct flight to the wintering grounds. The birds presumably become exhausted and fall into the sea, but their deaths are effectively due to starvation. It seems reasonable to assume that starvation (including starvation-induced predation or disease) is responsible for most of the natural losses of waterfowl.

6.4.3 *Predation*

Predation of eggs and young has been discussed above; excluding hunting by Man, predation of adult waterfowl has rather limited impact on populations. It represented only 0.14% of the natural mortality of waterfowl in Stoudt and Cornwell's (1976) survey. Most predation (70%) was by mammals and the remainder by birds and the vast majority of predated birds were dabbling ducks, with very few geese and swans.

It is not surprising that predation is a minor factor, since the birds are vulnerable for rather short periods, mainly during nesting and the moult. Predation on the nest can be a local problem, largely affecting incubating female ducks. This does not, however, make sufficient impact to unbalance the sex ratio in many species, since males are more susceptible to hunting

(Aldrich 1973). Geese are sometimes also taken on the nest by foxes and other canines, but such predation is of negligible importance on a population level (Owen 1980a, p. 173). Losses of adults during moult, although they do occur, are of negligible importance, as are losses to avian and other predators in winter. Avian predation on ducks is not well studied, though it occurs in wintering flocks of Teal in the Camargue, France, where birds are taken by marsh harriers (Tamisier 1974) and in Eiders in Norway, where they are taken by sea eagles. Overall, however, predation of adults is a trivial factor in waterfowl population dynamics.

6.4.4 *Man-induced mortality*

Direct killing for food and sport represents the most important influence of Man on quarry species. The annual kill of ducks in North America is in excess of 15 million and those in Europe 11 million. The respective figures for geese are 2.1 million and 200 000 (Scott 1982). The effect of this level of harvesting on populations is variable by species and complicated by the interaction between hunting and other sources of mortality. The control of numbers is discussed briefly in the final section of this chapter.

Mortality from indirect influences of Man can be serious in some circumstances and many examples are given in Chapter 7. Poisoning is one of the most significant of these; in the United States, poisoning of ducks by ingested lead pellets is considered so serious that lead is to be banned for all waterfowl hunting by 1991. Mallard, Pintail and diving ducks are worst affected; some populations are estimated to suffer up to 10% mortality from this cause alone (from figures in Bellrose 1959). Lead pellet densities in parts of Denmark were extremely high and lead is now banned there too. In the Camargue, France, the situation is potentially the most serious yet reported, with the excess of 60% of some species collected from wildfowlers' bags having ingested lead (D. Pain, pers. comm.).

Poisoning from the ingestion of lead weights from fishermen accounted for large losses of Mute Swans in England and resulted in the extinction of local populations (Goode 1981). Poisoning from agricultural chemicals and other forms of pollution can also give rise to serious local problems (see page 159).

Waterfowl have a high wing loading, which makes them relatively unmanoeuvrable in flight and vulnerable to aerial collisions. Such collisions were responsible for a substantial number of deaths of waterfowl in North America (Stoudt and Cornwell 1976). Swans are the heaviest and most vulnerable. In Britain about half the Mute Swans whose cause of death was known were killed after collisions, mainly with overhead wires. Even in this fully protected species, Man-induced mortality accounted for 85% of the reported deaths (Ogilvie 1967).

There is little direct evidence of the contribution of natural vs. Man-induced mortality; 'natural' mortality is usually calculated by difference between the overall mortality rate and the hunting kill rate. In the well-studied Mallard populations in North America, the kill by hunters (including an estimate for the unretrieved kill), accounts for around 45% of the annual mortality (Anderson and Burnham 1976). In geese, where for most populations numbers are being held down by hunting, natural mortality accounts for a very small proportion of deaths. In the protected population of Barnacle Geese breeding in Svalbard, natural mortality rates during the 1970s, as determined directly by the disappearance rate of marked birds, were around 6% annually (Owen 1982). As numbers increased further in the 1980s however, the natural rate rose to over 10% (Owen and Black 1990). Clearly the various sources of mortality interact with one another in a complex manner to determine the number of birds in a population.

6.4.5 *Mortality in relation to age and sex*

When there is competition for resources, it is usually young and inexperienced birds that suffer, and there is ample evidence of differential mortality with age in waterfowl. In the long-lived Mute Swan, the mortality rate of fledged young decreases from 59% in the first year of life, to 32% in the second and third and just over 20% as adults (Coleman and Minton 1980). Black Swans (which were legal quarry) banded as cygnets in New Zealand had an annual mortality rate of 31% in the first and second years, 20% in the third and fourth and 15% as adults (Williams 1973). Young birds are more vulnerable to hunting as well as natural mortality. This is due to inexperience as well as the need to disperse and travel further in search of food than adults established on traditional breeding and wintering areas. Ogilvie (1967) found that Mute Swans in their second year were more likely to die from collisions with overhead wires than were either younger or older swans, and the difference was highly significant. He attributed the difference to the fact that the 2-year-olds spent more time flying, at the time they were dispersing to find territories.

In hunted populations of geese, young birds are much more likely to be shot than adults; in Lesser Snow Geese, the annual mortality of first-year birds was 59% compared with only 25% for older geese (Boyd 1976). In protected populations the difference is much less marked; Owen (1982) found no consistent difference in mortality rates between adult and young Barnacle Geese in most years, though in a few years there was a significantly higher loss of young geese. In an almost unshot population of Canada Geese in Britain, young birds were only 1.05 times as likely as adults to die in a year (Thomas 1977). Yet, when conditions are hard, it is the young birds that suffer in competition with adults. In the cold winter of 1976–77 on the Atlantic coast of North America, the proportion of young Brant in a sample that died of starvation was very much greater than in the population as a whole (Kirby and Ferrigno 1980). Similarly, the proportion of young in the flocks of White-fronted Geese in Britain before the severe weather of 1962–63 was 13.3%, whereas later in the winter it had dropped to 4.7% (Boyd 1964a). In the Svalbard Barnacle Goose, mortality rate in the second year of life was significantly lower than either in the first or subsequent years (Owen 1982). This was attributed to the fact that second-year birds had gained in experience over young, but did not breed and so did not suffer the same stresses as adults. In hunted populations this difference is unlikely to be evident.

Differential vulnerability of adults and young is also very evident in ducks. Adult Mallard in North America have a mean annual mortality rate of around 42%, whereas that of young is 53%. Although young birds are more vulnerable to hunting than adults, the difference in the overall mortality rate is not only due to hunting losses; a similar proportion of the mortality of adults and young is attributable to hunting losses, i.e. young birds have a higher natural as well as hunting mortality rate (Anderson 1975). This means that young are more vulnerable to natural causes of mortality as well. Similar conclusions have been reached for a wide variety of species and populations of ducks in many different parts of the world.

There are also great differences in the mortality patterns of males and females in some populations and this has a great influence on the adult sex ratios of some species.

There are few data on natural mortality rates, which are usually swamped by those from hunting. Vaught and Kirsch (1966) suggested that females might be more liable to die natural causes because of the greater stresses they encountered in breeding (see page 30). In the protected population of Barnacle Geese, this was indeed found to be the case; the losses were suggested to be manifest on autumn migration (Owen 1982). A group of geese followed throughout their lives confirmed the relationship—the median lifespan of males was 10 years compared with 8 years for females (Owen and Black 1989b). The difference does lead to a slight excess of males in the adult population (Owen *et al.* 1978).

Both sexes of geese and swans are subjected to very similar conditions and near-equal mortality rates for the sexes is usually the rule. However, in shot populations, there is usually a greater likelihood for males to be shot than females. Imber (1968) determined the mortality rates in four populations of Canada Geese and found that males had consistently higher hunting mortality rates. In the population in New Zealand, males had a 31.2% annual rate, compared with 28.5% for females.

The pattern is radically different for ducks, where the sexes spend much of their lives apart and are subject to different pressures. The disparity in the sex ratios in flocks of wintering ducks first alerted scientists to the possible implications of differential migration and mortality patterns in North America (Hochbaum 1944) and in Europe (Lebret 1950). Rarely was it possible, however, to sample the whole range of a population and much of the difference was attributed to differential migration of the sexes (see Chapter 5). Owen and Dix (1986) calculated that the Wigeon population in Europe had 132 males/100 females and suggested a much higher level for Pochard. Detailed studies in North America identified some of the factors that may be involved. Competition for food between the sexes in northerly wintering areas and competitive exclusion of females has been found in the Canvasback (Nicholls and Haramis 1980) and other species of diving ducks (see e.g. Sayler and Afton (1981) for Goldeneye).

6.5 The control of numbers

Finally in this chapter we examine briefly how recruitment and mortality interact to control the numbers in a population. In particular, we consider on a population scale, whether a high mortality rate from one source lessens the rate of loss from another—**compensatory mortality**—or whether losses from different sources just add up to a cumulative total mortality rate for the population—**additive mortality**. The effect of different levels of hunting mortality on numbers under the two hypotheses is shown diagrammatically in Figure 6.5.

If we accept the hypotheses of Lack (1954), with which we started this chapter, we would expect all populations to be subject to density-dependent regulation. This is, however, very difficult to demonstrate and, in any case, in those parts of the world where populations are well studied,

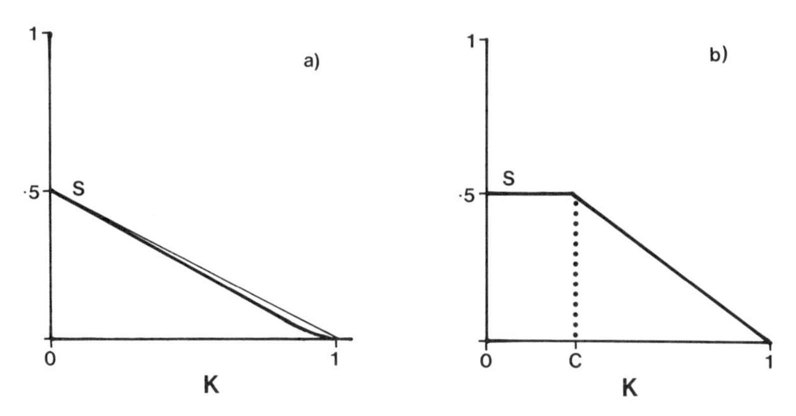

Figure 6.5 A diagrammatic representation of the hypothesis of a) completely additive mortality, where the annual survival rate, S, varies in relation to the hunting mortality rate, K. In b), mortality is completely compensatory up to a threshold level, C, beyond which the slope is proportional to K as in a). From Anderson and Burnham (1976).

they have been so modified by Man's impact on their numbers and habitats that they are unlikely to be behaving in the way in which they would without artificial influences. Knowledge of whèher mortality is additive or compensatory is essential to determine the impact of hunting on populations. As with most aspects of their ecology, geese and swans behave differently from ducks and examples are examined from each group.

We have argued that both in Europe and in North America, before Man cleared the forests, it was the availability of food in winter that set a limit on the size of goose populations (Chapters 2 and 6). Hunting pressure has increased in both continents in recent centuries however, and the winter habitat of those species that have taken to feeding on agricultural land has increased to the extent that it is, for all practical purposes, unlimited.

In the early parts of this century, there was concern that several populations of geese on both sides of the Atlantic were exploited to the extent that hunting was reducing numbers. Regulations were brought in to limit hunting and the populations in each case recovered. The example shown in Figure 6.6 is that of the population of Canada Geese in the Mississippi Valley Population. It shows that the population responded to the creation of a non-shooting refuge and to changes in the hunting kill. Thus, for this population mortality from hunting was additive and the level of exploitation controlled the numbers of birds in the population. This situation has been demonstrated for most populations of migratory geese and swans where information exists (see Ebbinge 1990 for review). In most

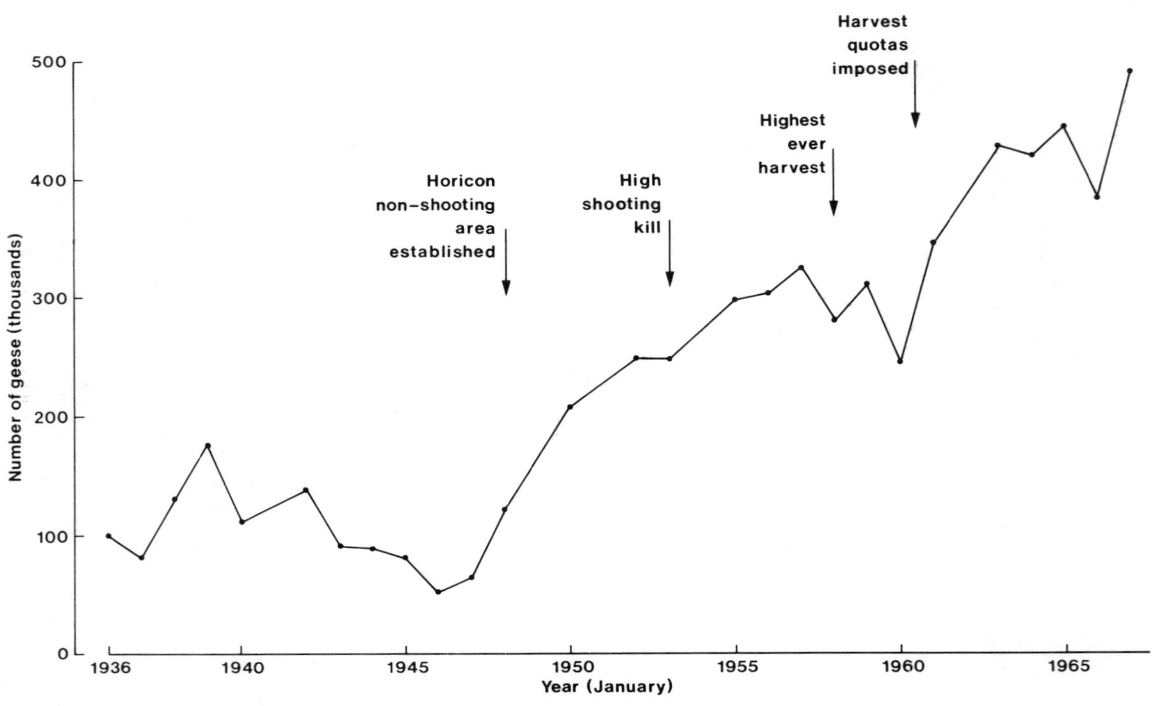

Figure 6.6 The growth of the Mississippi Valley Population of Canada Geese and some of the measures that were taken to control the mortality rate from hunting. Counts were in January, after the shooting season. Based on figures in Reeves *et al.* (1968), reproduced from Owen (1980a).

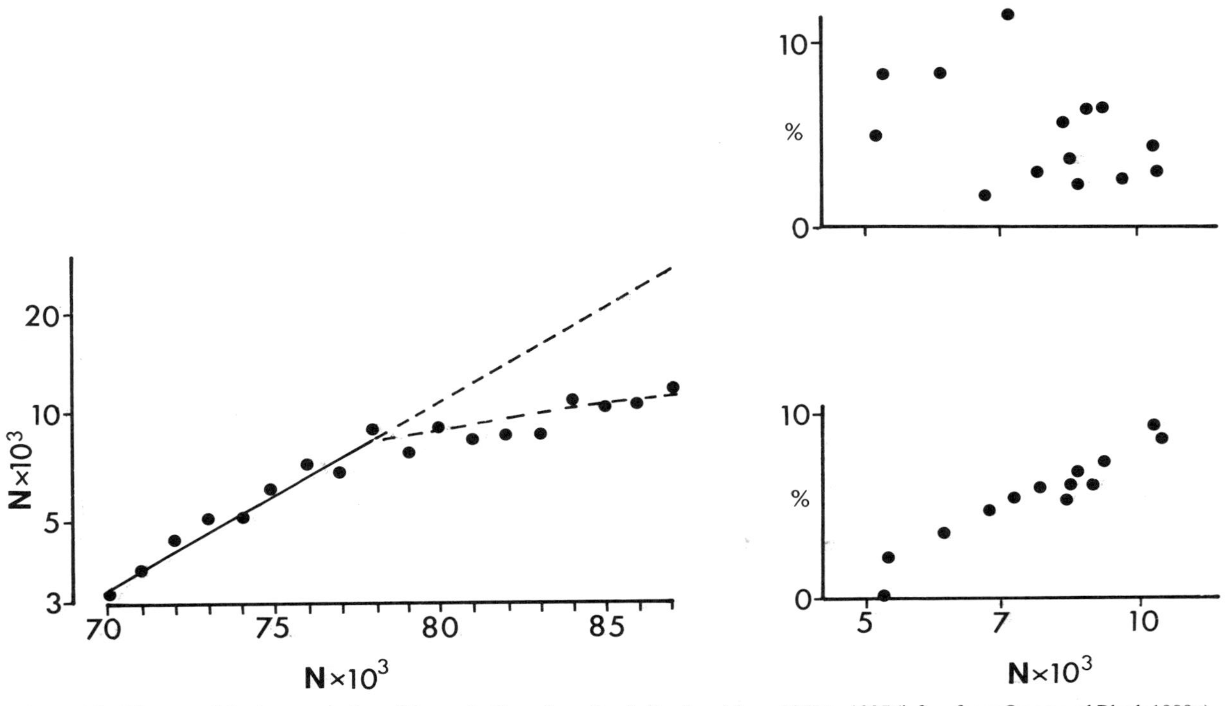

Figure 6.7 The growth in the population of Barnacle Geese breeding in Svalbard from 1970 to 1987 (left—from Owen and Black 1989a), and the mortality rate of adult females during (upper right) and outside (lower right) the shooting season in relation to numbers. Population size is plotted on a log scale.

cases, shooting has been re-imposed after the populations have recovered and few have grown to the extent that they have been density-regulated.

The one well-documented exception among natural migratory species is the Svalbard Barnacle Goose (Figure 6.7). The population grew rapidly following protection from shooting and the provision of refuges, from 300 in the late 1940s to 12 100 in 1988; this was due to a reduction in the mortality rate from shooting rather than to any change in the recruitment rate (Owen and Norderhaug 1977 and updated).

The figure shows the mortality rate of adults in the shooting season (largely illegal hunting kill) and outside it (largely natural mortality). There is clearly a very close relationship between the 'natural' mortality rate and the number in the population, suggesting density dependence. However, shooting-related losses show no density relationship; indeed, if anything there appears to be a negative correlation (though this is not significant). The exact mechanism of the density-dependent control is yet to be investigated, but there are strong suggestions that the limiting factor, which also operates to limit recruitment, is competition for food for breeding adults and young on the nesting and brood-rearing areas (Owen and Black 1989a). Similar limitations on recruitment in relation to density have been demonstrated in the Lesser Snow Goose colony at La Perouse Bay, Manitoba, though there is no suggestion that this is operating on a population scale (Cooch *et al.* 1989).

The situation we have just described could well be artificial, resulting from the abundant food supply in winter, but it seems likely that the same pattern will be found in other rapidly expanding populations. One of the effects might be to cause damage to the breeding habitat which will take many years to repair, as is happening in the La Perouse Bay Snow Goose colony.

A particularly interesting aspect of some populations is the inverse density dependence of hunting mortality. This is probably due to a 'predator-swamping' type phenomenon, the result of which is to make hunting an ineffective or inefficient method of limiting populations once they have grown beyond a certain size. This appears to have happened in the populations of Greater Snow Geese in North America (Gauvin and Reed 1987 and A. Reed pers. comm.), and Pink-footed and Greylag Geese in Britain (Fox *et al.* 1989).

The most detailed examination of mortality in ducks was carried out by Anderson and Burnham (1976), and has modified thinking and stimulated research into the subject. The hunting regulations in North America were designed to regulate the harvest and the overall kill, thus ensuring healthy breeding stocks for the following year. Anderson and Burnham demonstrated that mortality was compensatory for the Mallard, so that the number of birds shot in the autumn and early winter had no effect on the population in the following summer. They further suggested that compensatory mortality could well be the rule in duck populations. Clearly this only applies if hunting mortality stays below a certain threshold level. The threshold level depends on the reproductive characteristics of the species and the natural stresses to which it is exposed.

The threshold level has been estimated by Patterson (1979) as around 40% of the autumn population for the Mallard and only 10% for diving ducks such as Canvasbacks and Redheads. The levels of hunting in North America are well below these proposed threshold levels (Scott 1981), so that compensatory mortality is probably the rule. The intensity of duck hunting in Europe is, however, much higher (Scott 1982), so the assumption of compensatory mortality may be more problematical. In the Mallard, the only species studied in detail, density-dependent regulation, operating through overwinter loss, has been demonstrated (Hill 1984).

Nicholls *et al.* (1984) examined the evidence that had accumulated since the work of Anderson and Burnham for compensatory and additive mortality in North American duck populations. Their conclusions were that whereas there was conclusive evidence that mortality in Mallard (except perhaps for juvenile females—see Nichols and Hines 1983) was compensatory, but that the understanding of the processes regulating the populations of other species are little understood. For such species as the diving ducks, whose reproductive rates are much lower and which consequently have lower threshold levels, additive mortality effects at current harvest rates are a possibility.

Despite an enormous amount of research, especially in North America, our understanding of population dynamics of waterfowl is still poor, though intensive studies of geese and swans are making substantial contributions. In the absence of hard evidence for compensatory mortality, it seems that the conservative approach to hunting regulations, which assumes at least a level of additive effects of hunting, seems to be the safest way of proceeding in order to safeguard duck populations.

CHAPTER SEVEN
CONSERVATION AND MANAGEMENT

7.1 Introduction

This chapter examines the problems of waterfowl and wetland conservation and management. Because there is an interest in hunting the birds for food and sport, there is also an interest in safeguarding populations to provide future hunting opportunities. In population terms, conservation and management are more or less synonymous. In North America waterfowl are regarded as a 'natural resource' which is managed to ensure future harvests. In the Old World, species and habitats are conserved for their own sakes; the harvesting of ducks and geese is rather incidental. The two philosophies do, however, lead to the same kinds of activities being carried out, to ensure sizeable populations, to provide habitat and to minimise the effects of deleterious factors such as pollution and disease. First we consider the conflict between waterfowl and Man's commercial interests before looking into the conservation problems and ways of managing populations.

7.2 Conflicts with agriculture and fisheries

Damage by waterfowl to crops occurs on the largest scale in the prairie provinces of Canada and the northern United States. Once, the prairie pothole region of the United States alone covered more than 33 million hectares. By the 1960s, less than half of this remained, the rest had been drained for agriculture (Bellrose 1978). This has not only reduced the production of ducks from the region but has radically changed wetlands which were used in autumn by ducks and geese breeding further north, in the less developed pothole regions of Canada and in the Arctic.

The result of this reduction in habitat is that these populations nowadays leave their nesting ponds in autumn and feed on the arable land surrounding suitable roosts rather than gather in the once extensive wetlands. The agricultural practice of leaving cereal crops to ripen in a swath makes them even more vulnerable to damage by waterfowl, which can easily alight in the cut crop and strip grain from the lying stalks. The problem is extensive despite the fact that many areas around refuge roosts are specifically set aside or, in some cases, crops grown and sacrificed especially for waterfowl (Vaught and Kirsch 1966).

In Europe, damage by ducks is unusual, but there are increasing complaints against geese in all the countries in which they winter (see various accounts in Ruger 1985). The tradition of setting feeding areas aside for geese is not well developed in Europe and the farmers' problems are often exacerbated when roosts are protected and large concentrations develop. As populations of geese have increased in Europe and North America through the restriction of shooting practices and the provision of protected refuges, farmers have increasingly called for compensation payments or for goose populations to be controlled.

Problems are equally evident with Southern Hemisphere sheldgeese, especially in the Falkland Islands, where the resident Upland and Ruddy-headed Geese are accused of severe damage to the marginal grasslands (Summers and Dunnett 1984). Even the small population of Cape Barren Geese in Australia cause considerable conflict when they concentrate on small patches of intensively managed and irrigated pastures (Dorward *et al.* 1980).

In the tropics ducks, and even flamingos, can cause considerable damage, especially to ricefields. In Venezuela, irrigated ricefields provide an attraction for very large numbers of whistling ducks in the dry season. They do no damage on fields after harvest, but they also feed on newly planted rice or on unharvested grain (Bruzual and Bruzual 1983).

7.2.1 *Assessing damage*

Considerable efforts have been made in attempts to quantify the impact of geese on crop yields, either to justify culling or to convert losses into financial compensation payments. Losses of some crops are total and rather easy to quantify. In the prairies of Canada payments to farmers in some years have reached $1.5 m and the total cost of compensation and prevention together are up to $2.4 m. In some cases the federal and provincial governments buy a complete crop from a farmer and sacrifice it to the birds (Boyd 1980).

In other situations, especially on growing crops, the damage is more difficult to quantify. An example, from one of the most intensive recent

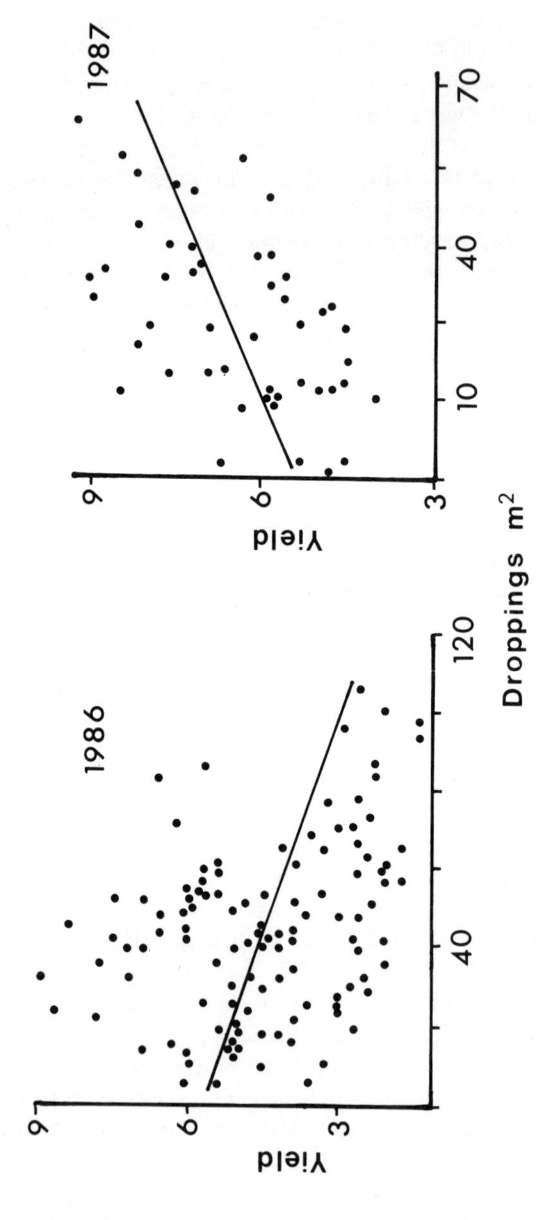

Figure 7.1 The relationship between grazing pressure by Barnacle Geese in winter and spring and the yield of silage at harvest in mid June on the island of Islay, Scotland in 1986 (left—correlation coefficient $r = 0.349$, $P < 0.001$) and 1987 (right—$r = 0.519$, $P < 0.001$). From Percival (1988).

Figure 7.2 On their spring staging areas on the St. Lawrence estuary, Quebec, Canada, Greater Snow Geese desert their mudflat feeding grounds to graze the lush pastures nearby. Note the stained heads of the geese—iron staining from the estuarine mud—and the juveniles, showing grey wing feathers, in centre foreground.

studies in Britain (Percival 1988), is given in Figure 7.1. In 1986, there was a significant depressive effect on silage yield in June, of goose grazing up to the end of April. The relationship was, however, not close; only 12% of the variation in silage yield was accounted for by goose grazing. In 1987, there was a significant **positive** effect of goose grazing, with 27% of the variance explained. This conflicting result was interpreted as being due to differential effects of weather. In the early (1987) season, the grazing stimulated growth of the crop.

Detailed studies have also been carried out in the Netherlands, which similarly showed that the relationship between goose use and yield loss was not very close and that there were complicating weather interactions. This made damage assessment very imprecise (Bruinderink 1987). In Canada, Greater Snow Geese caused significant reduction in the yield of hay, but the amount of variation in yield loss explained by goose use was variable, between 19 and 46% (Bedard *et al.* 1986).

Large numbers of studies have also been carried out on winter- and spring-sown cereals grazed by geese in the spring. Various methods have

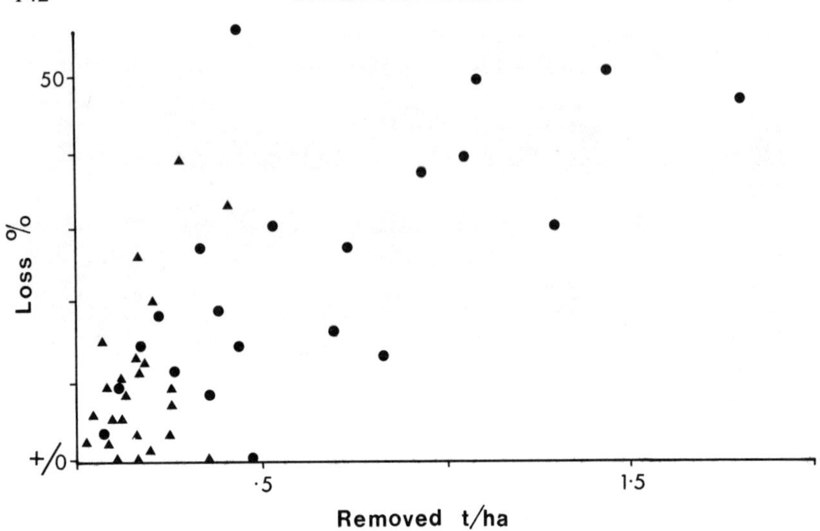

Figure 7.3 The relationship between the amount of material removed by geese (triangles) and by clipping (dots) and the loss of grain yield at harvest. Each point represents a separate trial and a number of different studies are included (plotted from figures given by Patterson 1990—sources of individual data quoted by him). Note that some of the points on 0 on the y axis indicated increases in yield after grazing in comparison with controls; correlation coefficient $r = 0.714$, $P < 0.001$.

been used, including excluding geese from the crop and comparing yields on grazed and ungrazed areas, and using penned geese at different intensities to graze the growing crop. The most common method, and the most easily controlled, is to simulate the action of the geese by clipping vegetation at various stages in the season. Patterson (1990) summarised a number of studies that had examined the effect of vegetation removal on the subsequent yield of grain. He calculated the removal rate from natural goose grazing as well as from clipping so that all the studies could be directly compared; his result is reproduced in Figure 7.3. There is a clear positive relationship between loss of yield of grain and the amount of material removed during the early growing season. The predictive precision of the trials is, however, poor; only 51% of the variability in yield is accounted for by the amount of material removed. Notice, also, that all the removal rates higher than 0.4 tonnes/ha refer to clipping experiments. The trials involving natural goose grazing are extremely inconclusive, indicating yield losses of up to 30% on the one hand, and on the other a positive effect on yield at rather low levels of grazing in some trials.

We conclude from this that it is impossible to operate a fair compensation system for goose damage to growing grass and cereals. Despite this some countries do operate such a system, using assessors to quantify damage after the event. In the Netherlands, money to pay farmers is obtained from a 'Game Fund' accumulated from hunting licences. The compensation payments have been almost up to $1 m (van Welie 1985).

7.2.2 *Preventing damage*

Substantial efforts have been made to develop effective methods of deterring geese by non-lethal scaring methods, but these are most successful only if there are undisturbed feeding areas nearby where the birds can feed unmolested. In practice, this often means the crops of a less vigilant neighbouring farmer. In any event, scaring is an intensive operation, only effective if it is continually monitored and changed in reaction to changes in the behaviour of the birds.

Another method of preventing damage would be substantially to reduce waterfowl populations. This may not be possible for the more numerous duck species, where hunting mortality is probably compensatory, but it would be for geese, where such mortality is additive to natural losses (see page 133). This is where crop protection comes into conflict with waterfowl conservation. Most of the populations of geese that cause damage are migratory, and breed, winter and stop on migration in the territories of different states. There is thus an international responsibility, enshrined in legislation such as the Migratory Birds Convention Act, agreed between the United States and Canada in 1918 and the 1979 European Community Directive on the Conservation of Wild Birds. Unilateral action by a single state can have widespread repercussions on the interests of another and, in Europe at least, there are widely differing traditions and practices on the exploitation and conservation of migratory birds.

The most realistic method of solving the problem of agricultural conflict thus seems to be to set aside refuge areas where the geese are tolerated and which enable neighbouring farmers to disturb birds to a protected area (Owen 1977b). These refuges are ideal for small populations with restricted distributions, whilst less intensive schemes may provide the solution for numerous and widespread species. Such schemes might involve the designation of rather large areas where geese are numerous, and where farmers can obtain payments for tolerating geese on their land. Outside these areas, restrictions on disturbance and shooting in defence of crops could be more lenient (Owen 1990). In the 1990s, when agricultural surpluses in Europe and North America are causing governments to urge the de-intensification of agriculture, such schemes may have some chance of being implemented.

7.2.3 *Problems for fisheries*

Waterfowl are also accused of causing damage to fisheries and these come into two main categories: the predation of molluscs from commercial shellfish beds and the predation of game fish from rivers and lakes. The problem of predation of molluscs is rather new and not well understood, but studies are proceeding on the impact of Eider Ducks on cultivated mussel beds in north-west Scotland.

The predation of Goosanders and Red-breasted Mergansers on salmon and trout fry on the other hand, has been the subject of intensive investigations since the 1930s (references in Wood 1987). The results are, however, difficult to interpret, since the predation takes place on small fry and smolt in the spawning and migrating streams and the 'harvest' by Man is some years later. It is evident, however, that sawbilled ducks can exert substantial predation pressure on the young of salmonid fishes. In the most recent study, Goosander broods on Vancouver Island removed quantities of salmon fry equivalent to 24–65% of smolt production assuming fry survival was density-independent (Wood 1986). If, however, density-dependent effects operated during the growth of the fry (as is likely), it is possible that the birds had no effect on smolt production, let alone on the production of salmon of harvestable size.

Figures indicating predation rates such as these will ensure that sawbills will continue to be persecuted and killed in large numbers by fisheries authorities, though their exact impact on stocks of adult fish may never be quantified.

7.3 The protection of waterfowl species

Again, we do not make the distinction here between the safeguarding of species or populations for hunting and their conservation for their own sake. Legislation and codes of practice, where they exist, all over the world are aimed at establishing and maintaining viable and sustainable populations. Species-related protective measures are most well developed in

North America, but they have been mimicked to a greater or lesser extent in other countries.

Waterfowl are large and provide a good source of food; they have been exploited by native peoples for millennia. The impact of Man on populations has, however, increased markedly in the last century, as firearms and ammunition have become more effective and the more isolated parts of the range have become accessible.

It was in North America, at the turn of the century, when there were no restraints on when or how many birds could be shot, and when many of the populations were being over-exploited, that legislation aimed at conserving stocks was first developed. No species had become extinct, but local populations of the more sedentary species such as the Giant Canada Goose had been wiped out and the Trumpeter Swan was severely threatened.

7.4 Hunting regulations

It was not until 1918 that the Migratory Bird Treaty Act implemented the treaty agreed between the United States and Britain, on behalf of Canada, two years earlier. The Act outlawed the commercial exploitation of migratory birds, making market hunting, which was practised widely, illegal. Spring shooting was also banned, a close season was imposed on all species and the Trumpeter Swan was given complete protection. A similar treaty was effected with Mexico in 1936. Since then an ever more sophisticated and sensitive system of harvest regulation has been adopted by or imposed on American hunters (Linduska 1972). These regulations and practices are described here to give some indication of how they might have an impact on waterfowl populations and their habitats.

7.4.1 *Hunting seasons*

The simplest and most common means of regulating shooting is to impose a limit on the period when certain species may be shot—the 'open season'. Since the birds migrate over long distances, clearly the times when they may legally be hunted have to be varied according to the availability of birds and their vulnerability at different times. The North American regulations recognise this by being organised on a flyway basis. Agreed seasons are imposed by a Flyway Council operating over the whole of the migratory corridors of waterfowl populations. The management of shooting seasons can, under this system be very specific, giving protection to those populations or species that need it, whilst allowing more liberal rules for more numerous ones.

A fundamental part of the operation of a system that is flexible and able to respond to changes in the fortunes of the birds is that it requires that the information on the numbers and breeding success of the quarry is available to managers and available before the season opens. In North America surveys are carried out each year in an attempt to quantify populations and to assess breeding success. More than 80% of the waterfowl wintering in the United States and Mexico breed in Canada, so international co-operation in such surveys is essential and the costs are shared. Clearly the accuracy of the population indicators depends on the species and the extent of their range. For example, Greater Snow Geese or Black Brant can be censused on staging areas on migration. Both species have easily identifiable juveniles, so that year's productivity can easily be assessed. The more numerous and widespread duck species are however, more difficult and their numbers are assessed by counts of females or pairs in the spring and production by brood counts after hatching (Dzubin 1969). The accuracy of these counts has been called into question and one study in British Columbia concluded that single surveys were inadequate to assess the number of breeding pairs or broods. The information was variable and inaccurate however, even after several surveys had been carried out in the same season and over the same areas, especially for dabbling ducks (Savard 1980). At best these surveys give a guide to the numbers of ducks to be expected in the 'fall flight'. The likely productivity in any season can, however, be predicted to some extent by the conditions on the breeding grounds before nesting (snow cover for Arctic regions and wetness of the ground in temperate nesting areas of ducks).

7.4.2 *Other limits*

There is a range of measures available to the waterfowl manager within the shooting season (Linduska 1972); the most common of these is the bag limit. This imposes a number of birds that can be shot, in a day or a season, by an individual hunter. The limit can be species specific, for example only one or two birds of a less common species may be taken even though the overall limit exceeds this. Bag limits can be imposed on a local scale, in a state or a county or on individual shooting areas. An extension of the bag limit is the **harvest quota**, where the allowable harvest is set and when this is reached the season is automatically closed.

An even more sophisticated method available to waterfowl managers is the **points system**, where individual species or sexes are allotted different numbers of points and set a limit on the number of points for individual hunters. For example, if a Mallard female is allotted 90 points and a male 10, and the total allowable points is 100, a hunter shooting only Mallard can shoot from two to ten ducks depending on the order in which they are shot. If the first bird to be shot is a female Mallard then the hunter is only allowed one more duck in the bag. He can legally shoot nine males and one female however, provided the female is the last to be shot. The system is clearly open to abuse and is more reliant than any other regulation on the honesty and good practice of hunters.

7.4.3 *Flexible regulations in action*

An interesting study of how different regulations worked in practice was carried out in Michigan (Mikula *et al.* 1972). Three different systems—a daily bag limit of two birds, a points system that was weighted in favour of particular species and of females, and a bag limit of four which could only include single ducks of a number of less numerous species. In a careful study, which avoided hunter and other sources of bias, the actual number of ducks shot, as determined by independent observers and not the hunters themselves, was compared under each system.

The regulations significantly affected the composition of the bag, and there were substantial differences between them. The highest kill rates of ducks as a whole were made under the two-duck bag system and the points system was intermediate. Hunters attempted, and to a large extent succeeded, in distinguishing between species and sexes in flight in order to maximise their bag. For example, the sex ratio of Mallard in the bag under the points system was 2.5 males per female, whereas it was only 1.3 males per female under the systems which did not favour females. The same applied to species which were highly rated. There were fewer violations of the points system than the others and hunters preferred it. The system does, therefore seem to succeed in its objectives of sensitive management of the harvest, maximisation of hunter performance and satisfaction under the system studied. It seems unlikely, however, that its performance would be as good under a less controlled and supervised regime.

The potential of a flexible system of hunting control is shown by the recovery of the Mississippi Valley population of Canada Geese (see Figure 6.6, page 134). Numbers of geese in the population used to be

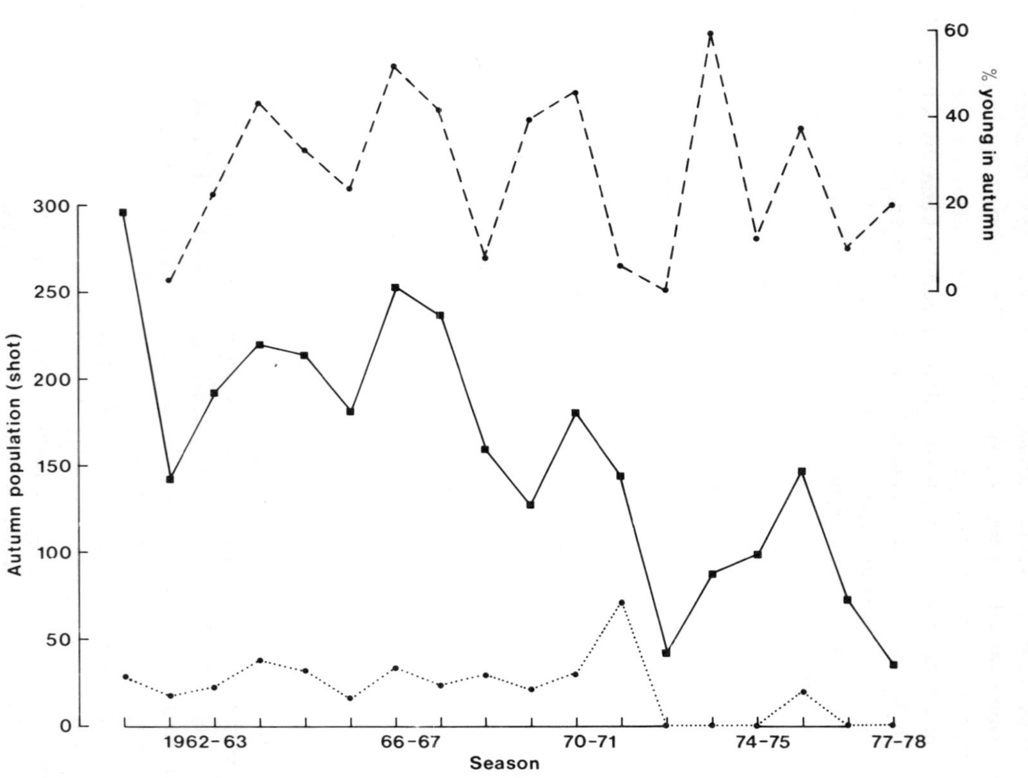

Figure 7.4 Numbers (solid line), breeding success (dashed line) and estimated harvest (dotted line) of Atlantic Brant between 1960 and 1977. Hunting seasons and bag limits were varied according to the size of the population. The most restrictive was a season of 10 days and a bag limit of 3, whereas the most liberal was a 70-day season and a bag limit of 6 birds per day. In 1972–1974 and in 1976–1977 there was no open season. Drawn from figures in Rogers (1979) and reproduced from Owen (1980a).

around 200 000 and the population used to winter on the Gulf coast. The creation of refuges further up the flyway made the geese more accessible to hunters for longer than was traditional and numbers had fallen to only 45 000 by 1945. Hunting was prohibited for one complete season and in following years a bag quota was set. As soon as the hunting quota was reached, the season was closed. Refuges were also created and the population grew to close to 500 000 in 1967. Regulations could then be more liberal, the population was in a healthy state and the size of the overall bag was very much greater.

There are limits to the sensitivity of shooting regulations; rules are always open to abuse, and natural factors can have a complicating effect. This is illustrated by the example of the Atlantic Brant population described by Rogers (1979) and shown in Figure 7.4. Following great declines in response to the die-back of the Brant's main food—*Zostera*—hunting was prohibited from 1933 to 1951. Controlled shooting was again allowed (a very restricted season of ten days and a bag limit of three birds) in 1951, and the situation closely monitored. Figure 7.4 covers a period of varying fortunes for the population.

The closed hunting period and the following restrictive harvest, coupled with a recovery of *Zostera*, allowed numbers to reach almost 300 000 by the early 1960s. Restrictions were then relaxed to allow shooting for 70 days with a bag limit of six birds—the most liberal allowed during the whole period covered by Figure 7.4. It is evident, however, that although the season and bag limits remained the same during the 1960s, the estimated harvest varied considerably. In 1971, food was scarce and the Brant had to forage in areas where they were more accessible to hunters. The effect was dramatic, with an estimated 71 000 (47% of the population) shot and a drop in numbers from 151 000 to only 73 000 in 1972. There followed a recovery so that a short hunting season was again allowed in 1976. The estimated kill was only 20 000 but numbers fell by over 50 000. The following year was a moderately good one for breeding, but in one of the coldest winters ever recorded, numbers fell from 115 000 to 52 000 between November and February. Numbers were even lower in the following year.

7.4.4 *Do flexible regulations work?*

The above example illustrates that even in one of the best documented populations and infinite flexibility in the control of seasons and bag limits, a population became in real danger. The whole system of complex regulations and expensive administration has been called into question by Matthews (1981), and the available evidence does suggest that the control of the harvest by the imposition of restrictions is far from perfect, as in the above example. Boyd (1983) however, argues that the necessity for close monitoring of populations imposed by these sensitive regulations brings its own rewards. At least it enables major changes in numbers or shifts of distribution to be detected. He cites in support of this view that all waterfowl populations in North America are in a healthy state despite a high level of hunter activity and success. Certainly in a number of cases prompt action has been taken to remedy possible serious situations, such as the Brant example and declines in waterfowl in Alaska; those populations did recover. Such detailed monitoring also provides a good basis for the analysis of causes, on which predictions for future prospects and threats can be based.

In other countries shooting regulations are generally less restrictive and less well controlled, though in some areas, such as New Zealand, a similar system operates. In most other countries legislation protects individual species (perhaps only in some regions), and allows a standard shooting season on quarry species. Variable-length seasons and bag limits are uncommon. Scott (1982) calculated that, under the *laissez-faire* system of regulation which operates in Europe, involving a long season over most of the range, 40–45% of the post-breeding populations of ducks falls to the gun compared with less than 20% in North America. Because ducks are four times as numerous in North America however, individual hunters shoot twice as many birds. Scott concluded that levels of exploitation in Europe were dangerously close to those that the populations could sustain.

Species can also be safeguarded by means of habitat protection and the creation of sanctuary areas where the birds can feed unmolested. In many ways a refuge system is more effective than protective legislation because it is more easily policed, especially with regard to regulating practices and ensuring that protected species are not exposed to the mortality and disturbance caused by shooting.

7.5 Threatened waterfowl

Waterfowl, apart from island endemics, are generally numerous, but a few species and populations are threatened and others have just been saved from extinction.

7.5.1 *Extinctions*

Four full species have become extinct—the Crested Shelduck, Pink-headed Duck, the Labrador Duck and the Auckland Islands Merganser (though some believe that a few individuals of the first two could still be at large).

The Crested Shelduck is known from only three specimens and its habits and reasons for extinction are completely unknown. The specimens were collected in far eastern USSR and South Korea, where a few individuals could possibly still exist (Collar and Andrew 1988).

The Pink-headed Duck was found in the forests of north-east India and was probably never numerous. A long-legged and long-necked species, it is thought to be intermediate between the *Anas* ducks and the pochards. The last confirmed report was in 1935, though the species was alive in captivity some years later. There is only a faint chance that small numbers might survive in Burma or Tibet, but in India their wet jungle habitat has all but disappeared. Habitat loss is said to have been the cause of their extinction (Ali 1960, Kear and Williams 1978).

The Labrador Duck was a sea duck of north-eastern North America; it was extinct by 1875, probably largely through egg-collecting and commercial exploitation for down (Sanderson 1978). Very little is known of its habits and few museum specimens remain but it is thought to have links with the Harlequin Duck and the Steller's Eider (Johnsgard 1965).

The Auckland Islands Merganser used to occur on mainland New Zealand, but was probably wiped out by Polynesian settlers and their introduced animals. The Auckland Islands, 500 km south of the southern part of the South Island was its last refuge in the latter part of the nineteenth century. The islands were inhabited from time to time, and rats, dogs and cats introduced. The species finally became extinct on its last refuge on the most southerly of the islands in 1902, when a number of individuals were shot (Kear and Scarlett 1970).

7.5.2 *Success stories*

Several local populations and subspecies have been wiped out by habitat loss, introduced predators and over-exploitation, yet there have been several celebrated cases of recovery from the brink of extinction. The Trumpeter Swan was numerous in north-central parts of North America until the middle of the nineteenth century, but by 1870, judging by the decline in the sales of swan skins, it was scarce (Banko 1960). By the beginning of this century it was thought that the species was on the verge of extinction; in 1932 there were only 69 known individuals throughout the United States. The southern population has now recovered following a determined programme of conservation, and other groups have been discovered breeding in Alaska. The species was never as rare as was feared and the Trumpeter Swan is now out of danger.

Perhaps the most notable case of the saving of a species is that of the Hawaiian Goose or Nene, which is described in detail in Kear and Berger (1980). This Hawaiian endemic, that was once numbered in many thousands, was reduced to less than 30 birds in the wild in the 1940s. The species no doubt descended from ancestral colonisers from North America, but is distinctive in being the most terrestrial of geese. The decline was due to the activities of settlers of Hawaii, either direct persecution for food or, perhaps more importantly, predation by introduced domestic and wild animals. The lowlands of the islands were settled and loss of habitat there, and competition with and predation by, feral animals, drove the Nene to the highlands of the volcanoes.

The plight of the Nene was recognised in Hawaii and elsewhere and a determined effort was launched to save it. The effort centred around rearing the species in captivity for release back into the wild, both in Hawaii and by Peter Scott at the Wildfowl Trust in England. The species bred well in captivity and large numbers of geese were reared and released into the wild. Unfortunately, efforts to study the birds in the wild were not as determined, and the ecology of the species and the real reasons for its decline (or even its subsequent revival) are little understood. It is, however, likely that the recovery of the Nene was due as much to the core wild population as to the released birds which, after generations of captive breeding and inbreeding, may not have been as well adapted to the wild environment (Kear and Berger 1980).

The population reached about 1500 birds in the 1970s (1761 Nene had been released by 1978), but despite continued releases, numbers have since declined to about 350 in 1990. More research on the ecology of the species in the wild is needed to ensure that the right measures can be taken to make the wild population self-sustaining. The intensive captive rearing programme did a lot to create greater awareness of the plight of the Nene and of waterfowl in general, but it has not yet succeeded entirely in securing the species in the wild. There is, however, little prospect of losing the species completely because of the ongoing reintroduction schemes and the many hundreds of Nene in captivity mainly in Europe and North America.

7.5.3 *Species still threatened*

The status of all the world's threatened birds has recently been reviewed by Collar and Andrew (1988); among them are 20 species of waterfowl (including the presumed extinct Crested Shelduck and Pink-headed Duck). Of these seven are endemics of island groups or a single island. The Laysan Teal, confined to a single island of only 285 ha, part of the Hawaiian group, nearly became extinct as a result of the introduction of rabbits and hunting in early parts of this century. The total population certainly reached single figures, but the population has now recovered, to fluctuate around about 500 birds. This may well be the carrying capacity of the habitat; no more than 700 have ever been counted (Kear and Williams 1978).

Two endemics of Madagascar, the Madagascar Teal, which is little known but very scarce, and the Madagascar Pochard, which has not been seen since 1970 despite searches, including an expedition in 1989, give great cause for concern (Collar and Andrew 1988).

Figure 7.5 The White-winged Wood Duck, probably the most endangered of the world's waterfowl. It is a rain forest species of the Far East; its remaining stronghold is thought to be in Indonesia, but there is little information on its status in the wild.

Among the non-island species, probably the rarest and most difficult to conserve is the White-winged Wood Duck—a forest perching duck of the Far East (Figure 7.5). It has suffered from deforestation and settlement in much of its range in Malaysia, India, Bangladesh, Burma and Thailand. The number in the wild is unknown but certainly small. The largest stock is thought to be in Sumatra, but its habitat there is being lost to agriculture (Collar and Andrew 1988). There are a number breeding in captivity and birds bred at the Wildfowl and Wetlands Trust have been sent to India and Thailand in order to start breeding stocks there with the intention of rearing birds for release to suitable habitats in the wild. Little is known of the behaviour or ecology of the White-winged Wood Duck in the wild; however, it is clear that the survival of the species depends on the protection of its rain forest habitat.

The White-headed Duck has disappeared from most of its range in Europe through habitat loss and persecution. On a world scale numbers are small, and the species is very vulnerale since about 70% of the population winters on a single lake—Burdur Golu—in Turkey where hunting from motor boats is common. The species now numbers around 12 000 and this number has fluctuated around this level for the past two decades (Anstey 1989). Reintroduction programmes are already in progress (see page 168).

7.6 The conservation of wetlands

Drained wetlands represent some of the most productive lands for agriculture, and in developed countries most wetlands have already been lost. As Third World countries progress, wetlands there are also becoming increasingly under threat, of drainage for agriculture and of having their feeder rivers dammed for irrigation and hydro-electric power. The protection of the remaining areas is vital to the future of waterfowl populations. The biggest challenge for conservationists is to safeguard the wetlands of developing countries, whilst allowing sustainable use by native peoples.

In Europe and Asia, wetlands have been drained and used for agriculture for centuries, but it is in North America that the pace of change has been greatest. It is there too that the remaining areas were first recognised as an internationally important resource. No doubt, the fact that waterfowl were considered valuable for food and sport, contributed to the protection of the wetlands on which they depend. The progress of wetland conservation in the United States and Canada has been reviewed by Sanderson (1978). The total area of natural wetlands in the United States is put at about 127 million acres (51 m ha) and in Canada wetland areas are counted in billions of hectares. By 1968 52 m acres (21 m ha—41%) of the wetlands of the United States had disappeared; in Canada much of the prairie pothole region has been drained but much of the vast wetlands of the north are still untouched.

With many waterfowl populations declining in the late nineteenth century, a start was made on using government money to buy and safeguard the key areas. The first National Wildlife Refuge was established in 1903; there are now some 400 covering some 80 million acres (32 m ha). In addition, state and provincial refuges and those owned and protected by non-governmental organisations make a substantial addition (Linduska 1982). From the outset refuges were considered to be vital to provide year-round habitats for migratory birds which travelled over great distances and need wetlands over their whole range. The migratory birds treaties between the United States, Canada and Mexico established a legislative framework.

The vast waterfowl stocks of North America visit only three countries, whereas elsewhere in the world, migratory flocks visit many different countries and their conservation becomes much more complex. Rivers crossing national boundaries and wetlands in one country might suffer from the actions of others. The International Waterfowl and Wetlands Research Bureau (IWRB) was established by the International Union for the Conservation of Nature in 1948 to concentrate on waterfowl and wetlands problems and to provide a scientific basis for their conservation.

In a historic meeting of IWRB in Ramsar, Iran, in January 1971 the final text of the Convention on Wetlands of International Importance Especially as Waterfowl Habitat (The Ramsar Convention) was drawn up. Criteria for establishing which wetlands were of international importance were established—mainly based on the numbers of waterfowl and other waterbirds they support. Any wetland holding 1% or more of a regional population or more than 20 000 waterfowl and waders is considered to be internationally important. Other sites, which do not meet the waterfowl criteria can be designated if the characteristics of the wetlands themselves deserve international recognition.

Each country which becomes a contracting party to the Ramsar Convention is required to designate a number of internationally important wetlands within its territory under the convention, which commits it to safeguarding the **ecological character** of designated wetlands. There is a proviso that in the **urgent national interest** a designated wetland can be destroyed, but in that event an equivalent area must be designated instead. The seriousness with which the convention is regarded is illustrated by the fact that no designated wetlands have been destroyed in the 15 years since the convention has been in force.

The convention was for its first years administered by IWRB, but in 1988 a Secretariat was established, with separate finance, based jointly in Slimbridge, England and in Gland, Switzerland. The first contracting parties joined the Ramsar Convention in 1974, but it soon gained momentum. By April 1989, 52 countries had become contracting parties, and 426 wetlands, covering more than 28 million hectares, had been designated. These ranged from Malta, which has designated a single site covering 6 ha, to Canada, with 30 designated wetlands covering 13 million hectares.

In many regions the value and importance of wetlands are not well understood and in recent years IWRB has been trying to fill in the gaps in knowledge. In 1987, the Directory of Neotropical Wetlands was published, covering the most important areas of South America (Scott and Carbonell 1986). A Directory of Asian Wetlands has just been completed (Scott 1989).

Progress has been made in recent years to convince people of the value of wetlands, and their importance, to people as well as wildlife. Wetlands are vital to Man as reservoirs for water, as protection against floods and as providers of food. The IWRB emphasises these human values, recognising that in the end it is the value of wetlands to humans that will decide whether wetlands will survive.

7.7 Remaining threats

Despite recent efforts there are still a large number of powerful threats to waterfowl and their habitats. In developing countries, drainage for agriculture, especially ricefields, is still a major source of wetland loss.

7.7.1 *Dams and developments*

The diversion of feeder rivers for irrigation schemes threatens some important wetlands. One of the most important wintering areas of Palearctic waterfowl—the great lake of Ischkeul in northern Tunisia—is threatened with just such a development. The 12 000 ha lake is a shallow freshwater one fed by several rivers. Dams are proposed on several of these rivers to supply water for irrigation schemes; two dams have already been built. The lake is connected to the sea via another, saline lake and when water levels are low in Ischkeul in summer there are incursions of saline water. The diversion of more water from the feeder rivers could have disastrous consequences, involving salination of the lake, which would change its character completely, and certainly make it less attractive to waterfowl. The fisheries of the lake under the present regime are of great economic importance, and studies, initiated by IWRB, are being carried out to try and reconcile the needs of agriculture and the maintenance of the lake.

Elsewhere in Africa, human activities such as burning and over-grazing by domestic stock are causing bush encroachment into wetlands in many areas, and in the temperate areas of southern Africa wetlands are disappearing rapidly (Siegfried 1970). The great inland delta of the Niger in Mali extends during the wet season to 2 million hectares (Scott 1980). It is vital not only for over a million waterfowl originating from the Palearctic as well as Africa, but also to the local population which depend on the delta for grazing for their stock and for fishing. There are no protected areas and the delta is threatened by degradation and over-grazing. Here again the key to the answer is that the maintenance of its value for birds depends on the sustainable use of the area by humans.

In Europe, where most of the original wetlands have already disappeared, threats continue. In Iceland many of the valleys which are important habitats for waterfowl, notably Pink-footed Geese, are threatened with flooding to provide water for hydro-electric schemes. Parts of the highlands are already under water and further dams have been approved.

In Britain, developments for industry and tourism threaten most of our estuaries. The use of power from the tides has been used for centuries in a small way, but there has only been one major scheme—at La Rance in northern France. Now several areas are being proposed for large-scale tidal power schemes. The Severn Estuary in the south-west of Britain has the second largest tidal range in the world and a 15 km barrage is proposed across its mouth. The power generated by such a barrage would, at peak generation, produce more than 5% of Britain's energy, so the scheme is very attractive.The main threat to waterfowl and wading birds is the possible loss of intertidal habitat as the mean tide level is raised. There have been two stages of environmental studies, but full feasibility studies have not yet been commissioned.

Several other developments threaten our estuaries—in South Wales, two barrage schemes, across the mouth of Cardiff Bay and across part of the Loughor estuary, are proposed to create lakes around which there are plans for substantial housing and tourism developments. Mudflats, regarded by many as unsightly, are attractive 'unproductive' areas for waste dumping and many estuaries are continually being eroded by such developments around the edges.

Developments further inland also threaten coastal wetlands. The creation of dams across the Mississippi River and the straightening of its course near the sea has stopped the massive amounts of silt that it carried from reaching the delta at its mouth. The result is that the delta marshes are being eroded at an alarming rate. Deforestation of the catchment also results in damage on the coast. Rapid run-off of rainwater erodes the soil and carries the silt downstream, causing wetlands to silt up and smothering marine life in the shallows around the mouths of rivers.

7.7.2 *Pollution*

Pollution from various sources poses severe threats to wetlands. Run-off of agricultural fertilisers causes eutrophication which results either in algal blooms or to increased growth of emergent vegetation which leads in rapid succession to terrestrial communities. Pesticides escaping into water-courses are toxic and have a direct effect on animal and plant life. The use of persistent DDT in the 1960s led to high concentration in fish-eating birds and caused egg-shell thinning and a reduction in breeding success (Cook 1973).

Oil pollution poses a direct threat to waterfowl, especially in shallow seas. The deaths of sea ducks from oil pollution in the Danish part of the Baltic was well documented in the early 1970s. Thousands of ducks were killed in major incidents each year and many others succumbed because of minor spillages and illegal tank-cleaning operations. In 1972 The total number of deaths from oil in Danish waters was estimated at 70–100 000 sea ducks (Joensen 1976). Possibly the worst incident ever occurred on 24 March 1989 when the Exxon Valdez struck a reef in Prince William Sound in south-eastern Alaska. The tanker spilt 232 000 barrels of oil into the sound and the slick, moved by strong winds, at one time stretched over 150 km. The effect on the sensitive Arctic ecosystems and their wildlife was devastating, and waterfowl and sea mammals may not be able to use the shallow water habitats and coastal marshes surrounding the sound for many years.

The sea is threatened as never before from effluents; the Baltic is already heavily polluted and the North Sea is giving cause for concern. The greatest threats are from waste dumping at sea and poisonous effluents which continue to be discharged into our rivers. On the other side of Britain low level radionucleides are being discharged into the Irish Sea and in affected areas shellfish are not safe for consumption. This degradation is not likely to be tackled until it begins to affect the human population. The algal blooms which devastated the Adriatic tourist resorts in 1988 and 1989 was a dramatic illustration of the effect of sewage pollution in the Mediterranean.

Poisoning from the ingestion of lead pellets originating from shotgun cartridges or from fishing weights is a common problem. In the United States, the problem was recognised in the 1930s (Shillinger and Cottam 1937). In some studies and species, the incidence of lead in the gizzards of shot ducks was as high as 15–20%, and a single pellet is sufficient to cause death. It is recognised as being serious enough to cause some states to ban

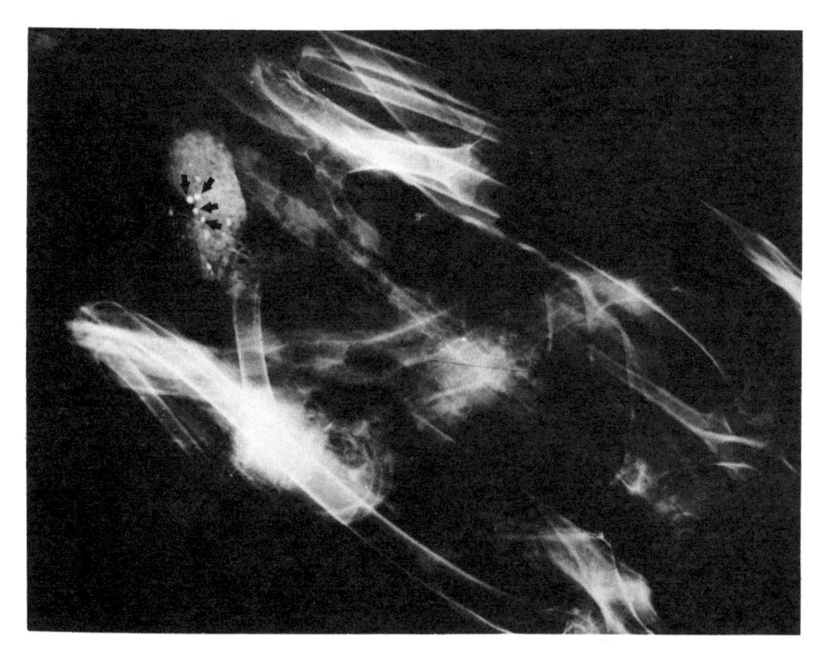

Figure 7.6 An X-ray photograph of a Bewick's Swan. The grey area at top left is the gizzard, which is full of grit. Note the four lead pellets (arrowed). The lower two have been heavily eroded. The swan died of lead poisoning.

the use of lead for waterfowl hunting, and the use of non-toxic shot will be required by law in all states by 1991.

In Europe problems of lead poisoning have been more localised. A study of the incidence of lead in soils indicated between 100 and 1800 pellets/m^2 in the top layers in heavily shot areas (Meltofte 1978). This has led to the banning of lead in waterfowl shooting in Denmark. The most serious lead poisoning in Europe caused the disappearance of local populations of Mute Swans from some lowland rivers in England. The ingestion of lead fishing weights was identified as the cause. On a national level some 3500 Mute Swans, out of a population of about 20 000 were estimated to die each year from lead poisoning in the 1970s (Goode 1981). This led the British Government to ban the sale of lead fishing weights in 1987. The swan population is showing signs of recovery despite the fact that there is still widespread use of lead in fishing (Sears 1988).

The provision of facilities for tourism and recreation not only threatens waterfowl through habitat loss, but causes disturbance to the birds and effectively reduces the carrying capacity of habitats. An intensive study of the effect of water-borne recreation on an inland lake in South Wales, indicated that disturbance from recreation had an impact on all species, and that the carrying capacity of the lake was held well below its potential (Tuite *et al.* 1983). Activities such as bait-digging and walking reduce the use waterfowl can make of estuarine mudflats, as shown in Figure 7.7. The intensity of use of the more disturbed *Zostera* bed is very much less than that of the undisturbed area. The *Zostera* is depleted in the autumn from natural die-back and the action of the waves, so the recreational pressure effectively reduces the food supply for Brent Geese.

As we have described, the effects of shooting on populations is less severe than in the past, but populations are still over-hunted in some areas. Protected species are also still shot. In Australia, there is widespread shooting of the threatened Freckled Duck despite the fact that it is in theory protected. The incidence of lead shot in the tissues of a range of protected and quarry species examined by the Wildfowl and Wetlands Trust differed little whether or not the species was protected. Clearly much education is required to convince hunters of the need for conservation of some species and to ensure that protection is effective in practice.

7.8 Creative conservation

This section deals with activities which are aimed at enhancing conservation in an active way. This can be achieved by increasing the carrying

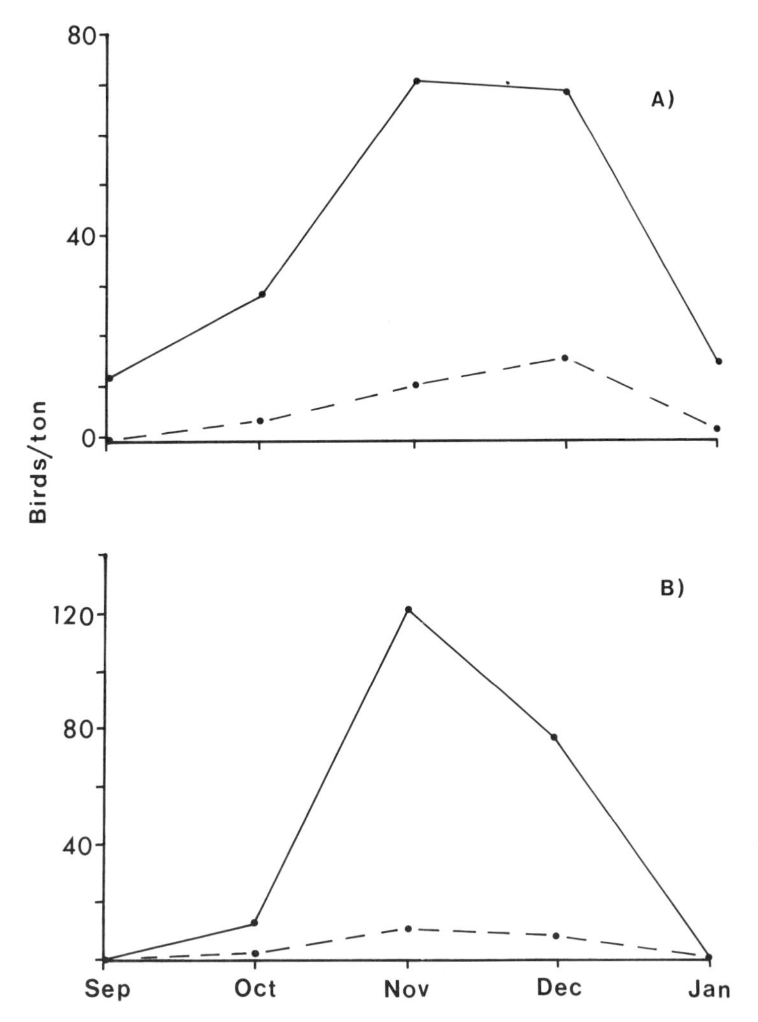

Figure 7.7 The number of Wigeon (upper) and Brent Geese (lower) using two mudflats in the Exe estuary per tonne of their food plant, *Zostera*. One area is undisturbed (solid line) and the other disturbed (dashed line). From D.V. Bell (unpublished).

capacity of habitats, allowing larger numbers of birds to be accommodated in a given area; by creating new habitats or re-creating lost wetlands; or by supplementing wild populations, either by translocating birds from one area to another or by releasing into the wild, birds reared in protected areas or in captivity.

7.7.1 *Increasing carrying capacity*

There are a wide range of methods that have been developed to maximise the value of wetlands for waterfowl and to increase carrying capacity. Here we focus only on a few examples to illustrate the principles and techniques involved. The concept of carrying capacity involves the idea that in nature, animal and plant populations will find an equilibrium level which is set by the amount of a limiting resource.

In his classic work, Lack (1954) argued that in most situations it is food that is limiting, though in special cases other resources such as nest sites can reduce population size. The resource is critically limiting only at certain stages of the life cycle. For example, in Chapter 2 we argued that most waterfowl populations were limited by the food supply in winter. It follows that if a resource 'bottleneck' is removed, the population will, under natural conditions, increase in size until the next resource becomes limiting. Normally populations show their greatest mortality at the time when they are at the limit of the resources. One of Lack's most important conclusions, which has been amply supported by later work, was that animals do not limit their breeding effort to fit resources for their young, but that each individual maximises its own breeding output. This leads to production beyond the capacity of the limiting resource. The important implication of this for waterfowl or any population which is being artificially cropped, is that there may be, at times before the limiting factor has begun to operate, a 'surplus' of animals which can be killed without detriment to the population (see Chapter 6 for discussion of compensatory mortality).

The principle of increasing carrying capacity is, therefore, to identify a limiting resource and supplement it to allow numbers to increase. One of the simplest ways of doing this is to provide nesting sites where there are few but where other conditions are suitable. Several species of ducks nest in elevated cavities in trees, but cavities are uncommon in new or managed forests where there are few mature or dead trees. The most spectacular example of how a population of hole-nesting species can be enhanced artificially is that of the Carolina or Wood Duck in the eastern United States.

The Wood Duck had been over-exploited in the late nineteenth century and numbers were very low when the Migratory Bird Treaty Act was passed in 1918. The species was completely protected until 1941. The shortage of natural cavities in the hardwood forests where the species nested was also identified as a probable limiting factor on breeding density. There followed an intensive programme to provide nesting boxes in places where natural cavities (usually excavated by the larger species of woodpecker) were scarce. The birds readily adopted the boxes and the nesting density increased well beyond what would have been possible in natural conditions (review in Bellrose 1978).

The use of boxes had another beneficial effect on production of Wood Ducks—nesting success was better than in natural cavities. The success of nests in natural cavities (a successful nest is one in which at least one egg hatches) in six studies ranged from 7% to 55% (mean 43%, n = 328). In 22 separate studies of Wood Ducks nesting in boxes (which are designed to repel predators) success rates varied between 32% and 95%, with a mean of 70% of a very large sample of 44 824 nest boxes (calculated from figures in Bellrose 1978). Since ducks return to nest in sites where they have been successful, this leads to a high degree of preference for boxes over natural cavities. In one study in Mississippi, none of 35 apparently suitable natural cavities was used by Wood Ducks, although 580 nests were found in boxes. The occupancy rate of the boxes provided was 90% (Strange *et al.* 1971). The nesting density of many hole-nesting species has been increased in this way.

The provision of artificial islands, either by excavation of moats, by dumping spoil or by creating floating platforms, has also proved of great value in increasing nest density and success. In southern Alberta, the density of duck and goose nests increased with the age of an island and the development of vegetative cover. Because the amount of water/land edge was more important than area *per se*, small islands tended to be more productive than large ones. Those far from the shore were also favoured and supported higher nest success, probably because they deterred mammalian predators (Giroux 1981). The control of predation is clearly important to nesting ducks, and artificial predator reduction in nesting areas allows high nesting density and success (Duebbert and Lokemoen 1980). In some parts of the North American prairies, there has been such a reduction in the area of nesting habitat that density on the remaining patches is artificially high. Predators find it easy to locate nests in such an area and can either be artificially controlled or excluded by fencing.

The quantity of food for wintering ducks depends on the summer management of marshes and the availability of that food on the water level in winter. In North America, impoundments are managed for waterfowl by varying the water levels. Drawdown of the water in summer replicates the natural drying out of water bodies in summer, but controlled drawdown and re-flooding can be used to maximise the effect in increased productivity. Annual plants colonise the exposed area, providing an abundance

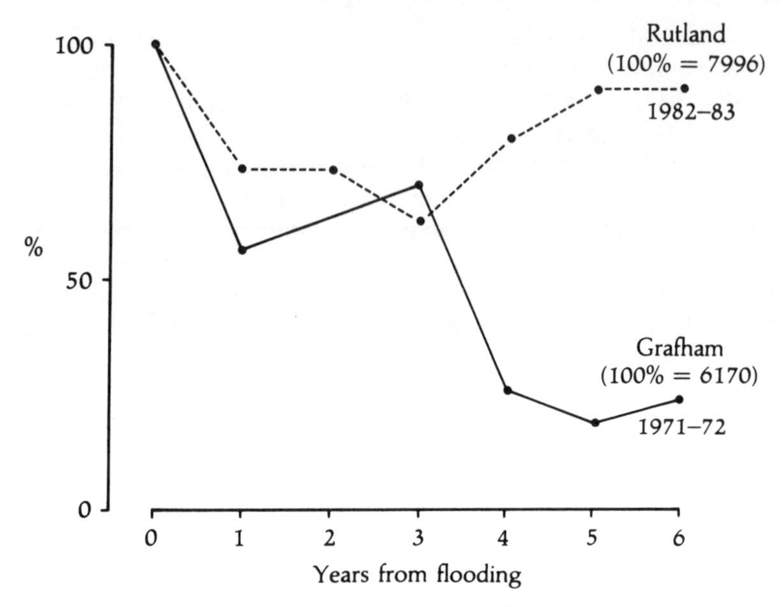

Figure 7.8 The numbers of waterfowl counted on Grafham Water, England from the time it was first flooded in 1965, and Rutland Water, first flooded in 1976, expressed as a percentage of the peak population. In Rutland, parts of the reservoir were designed to maximise its value for waterfowl. Reproduced from Owen *et al.* (1986).

of seeds, and soil fertility is improved (Kadlec 1962). The value of reservoirs to waterfowl tends to be greatest just after flooding, and stabilises in a few years to a level considerably lower than the early peaks. The 1976 drought in Britain caused severe drawdown of reservoirs, with the result that duck populations were high in the following winter. Dabbling ducks, especially the seed-eating Mallard and Teal, were the main beneficiaries. The pattern of numbers in the following years mirrored the decline after first flooding. Figure 7.8 illustrates how duck populations can be maintained at high levels by the creation and management of refuge areas.

In undisturbed areas, the quality of the vegetation is the most important characteristic influencing the choice of feeding area for geese (see Chapter 2). Food quality can be manipulated by management of grazing and cutting regimes during the grass growing season. Experimental management by cutting and fertilising semi-natural vegetation resulted in an increase in subsequent use of the treated areas by 87% over untreated controls (Owen 1975).

One of the consequences of successful refuge management is that unnatural concentrations of birds will gather, and this is not without its potential drawbacks, as described by Owen (1980b). One of the potentially most serious is the risk of disease epidemics. Botulism is one of the most serious diseases, and there have been outbreaks in many parts of the world, most associated with unnatural concentrations of birds (see page 127).

Concentrating birds may increase the adoption rate of novel feeding habits, and increase rapidly the prevalence of agricultural damage. Artificial feeding of waterfowl with grain in winter has been held responsible for the rapid adoption of grain-feeding in fields. Refuges can also cause distributional shifts in waterfowl, which are sometimes regarded as undesirable. This has happened extensively in the United States, where northerly refuges have allowed populations of waterfowl, especially geese, to stop short of their traditional wintering grounds, to the consternation of southerly hunters (see e.g. Dzubin *et al.* 1975).

7.8.2 *Habitat creation*

As natural wetlands have disappeared in developed countries, those created artificially by Man become increasingly valuable. The most important are reservoirs and wet mineral workings, chiefly gravel pits. Waters need not provide feeding areas as long as food is available nearby. The amount of agricultural habitats available for wintering Greylag and Pink-footed Geese in Scotland was increased considerably by the provision of inland roosts in the form of reservoirs (Newton *et al.* 1973). The birds were previously restricted to coastal and estuarine roosts and their margins. In North America and some parts of Europe it is common practice to flood areas in winter to provide roosts for waterfowl which feed on surrounding agricultural fields.

The increased area of water-filled gravel pits has made a substantial contribution to breeding as well as wintering waterfowl. The number of Tufted Ducks wintering in Britain doubled between 1960 and 1980, and the number in September (the local breeding stock and their young) trebled. More than 60% of Tufted Ducks in Britain as a whole are accommodated on artificial habitats. The proportion of six duck species and the Canada Goose on artificial, newly created waters, exceeds 40% of the total number in Britain (Owen *et al.* 1986).

The success of a newly created habitat designed and modified with the needs of waterfowl in mind, as well as sensitive management to reduce disturbance and maximise feeding opportunity, is illustrated by the

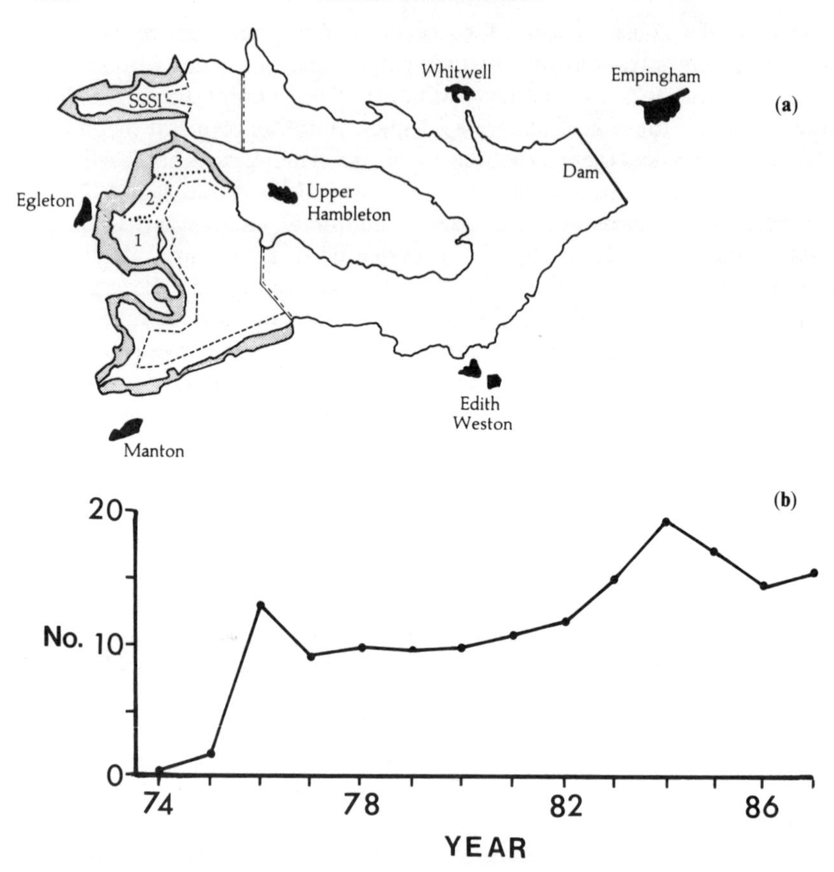

Figure 7.9 (a) A map of Rutland Water, England, showing the various zones and reserves designed for multi-use——— 50 metre boat limit, . . . lagoon bunds, - - - - sailing limit, shaded area shows reserve; (b) the number of waterfowl (the sum of the annual peak number of each species) counted on the reservoir from the time of first flooding in 1975. Note that in the first year the area of water was rather small. Based on Appleton (1982 and updated), map reproduced from Owen *et al.* 1986.

example of Rutland Water in eastern England (Figure 7.9). Before the reservoir was created, detailed discussions took place between the water authority and conservationists, and the needs of waterfowl and other wildlife were taken fully into account in the design of the marginal areas where reserves were created for wildlife. These reserves were transferred to a conservation trust and a warden appointed before a drop of water entered the reservoir. The number of waterfowl using the reservoir was maintained at very high levels in the years after flooding. This pattern contrasts with that in the nearby Grafham Water in the years before drawdown (Figure 7.8). At Rutland the decline in productivity of the main reservoir as it matured was counterbalanced by an increase in the productivity of the managed reserve areas. The diversity of wildfowl increased as the range of habitats developed and breeding populations of several species became established (Appleton 1982 updated).

7.9 Restocking and reintroduction

This section deals only with the supplementing of wild populations (restocking), and the reintroduction of birds into an area where they once were, and where they were eliminated, usually by Man's activity. True introduction of exotic species is undesirable (though vary many waterfowl have become naturalised in countries where they were not native) and has generally had no beneficial effects.

The motive of most of the restocking programmes that have been undertaken is generally to increase hunting opportunities. In Europe and in North America it is common to release Mallards, either bred on game farms or reared from eggs collected from the wild onto managed ponds for shooting. Many of these are shot early in the shooting season, but others survive and integrate into the wild Mallard population and some migrate with the wild flocks to distant breeding areas (e.g.Harrison and Wardell 1970). However, there is little evidence that release programmes increase the size of the wild Mallard population for following seasons. Indeed, if the size of Mallard populations is governed by the available food supply rather than hunting, as seems likely (Chapter 6), released birds will have no effect beyond midwinter, even though they might show comparable survival rates to wild Mallard (Lee and Kruse 1973). Only by providing new habitats can numbers be increased in the long term.

Birds reared in captivity are commonly released, or transported from areas where they are numerous, to extend the range of a species or to replace local populations which have been lost. The most extensive programmes have involved mass translocation of Canada Geese within the United States. Very large races of Canada Geese (Giant Canadas) used to be resident in most of the states, but many local populations were wiped out by over-hunting around the turn of the century. Geese were either trapped and transported to be reintroduced, usually in a protected area, or goslings reared in captivity were released. Usually the birds were grounded for a time by pulling their primary feathers, and supplementary food also given for a limited period. The success of these programmes was mixed. Many new flocks were established, but others failed. An attempt to establish Canada Goose flocks in the south-eastern states involved the trapping of over 20 000 geese further north and relocating them. The programme was an almost complete failure because few of the birds returned in following years (Hankla 1968).

Reintroduction using birds bred in captivity is widely advocated as a means of re-establishing populations which are close to extinction, but the track record of such projects is not encouraging. The case of the Nene has already been mentioned (page 152). Unfortunately the monitoring of that release programme was not good enough for us to know whether it was released birds or other measures taken to protect the wild stock that were responsible for the upturn in numbers and the present wild population is still not self-sustaining.

A more recent, and better documented case concerns the Aleutian Canada Goose, a distinct race breeding exclusively on the Aleutian Islands chain. The population was at one time reduced to around 300 birds in the 1960s and a captive rearing programme was begun. Birds were released in several islands but there were severe problems, especially with the predation of the flightless released birds by Bald Eagles *Haliaeetus leucocephalus*. It was not until shooting was stopped in the most important areas in the central valley of California however, that the population began to recover, to 1700 in 1977 (Woolington *et al.* 1979). Numbers have continued to increase and it is now safe, but not as a result of the reintroduction programme despite the large amount of effort and money that was devoted to it.

There are a few other active or proposed reintroduction programmes involving threatened species, such as the reintroduction of the White-headed Duck to Hungary and Sardinia. This is an expansion of present range to areas where the species occurred in the past and where the habitat is still available and protected. The avicultural performance of the Hungarian project has been impressive, and 180 ducks have been released into the wild since the project started. However, the fate of most of the released birds is unknown; only a handful have been resighted. Only with detailed monitoring of the fate of birds following release can the success of a reintroduction programme be properly evaluated.

Reintroduction programmes may have their part to play, but they must be carefully planned and followed up by proper monitoring. An international meeting on bird reintroductions in 1988 highlighted the problems and pitfalls and made the following recommendations:

1) The first step is to assess numbers and distribution and identify the remaining habitat and if necessary re-create or enhance it—there must be sufficient suitable sites available.
2) There should be international agreement and collaboration on projects involving mobile or migratory species.
3) Releases should be preceded by education and public awareness programmes.
4) The genetic and behavioural quality of the released stock should be the best available.
5) The releases should be gradually and carefully done and the performance of the birds monitored over the extended period.
6) There should be continual re-assessment and conservation of the birds and their habitat; no reintroduction scheme can be considered a success until it results in a self-sustaining population in the wild.

7.10 Conclusion

The story of waterfowl conservation is not a pessimistic one. Most stocks are in good shape and are well monitored so that signs of problems can be detected early. The group is fortunate that there are powerful vested interests in maintaining waterfowl stocks for shooting—they are a valuable resource. There are, however, a number of species which are highly threatened and should be investigated urgently. A number of others, because they are resident on one or a few small islands or rely on a narrow range of natural habitats, are likely to remain vulnerable.

APPENDIX

SCIENTIFIC NAMES OF SPECIES REFERRED TO IN THE TEXT

(The letters or numbers after the names of geese refer to Figures 3.1 and 3.8)

Common name	**Scientific name**
African Black Duck	*Anas sparsa*
Auckland Islands Merganser	*Mergus australis*
Australian Black Duck	*Anas poecilorhyncha*
Australian Grey Teal	*Anas gibberifrons gracilis*
Australian White-eye or Hardhead	*Aythya australis*
Bar-headed Goose (L)	*Anser indicus*
Barnacle Goose (Z)	*Branta leucopsis*
Barrow's Goldeneye	*Bucephala islandica*
Bean Goose	*Anser fabalis*
Russian Beam Goose (C)	*Answer fabalis rossicus*
Western Bean Goose (B)	*Anser fabalis fabalis*
Bewick's Swan	*Cygnus columbianus bewickii*
Black Duck (North American)	*Anas rubripes*
Black Swan	*Cygnus atratus*
Black-headed Duck	*Heteronetta atricapilla*
Blue Duck	*Hymenolaimus malacorhynchus*
Blue-winged Teal	*Anas discors*
Brent Goose (Brant)	*Branta bernicla*
Dark-bellied Brent (1)	*Branta bernicla bernicla*
Black Brant (2)	*Branta bernicla nigricans*
Atlantic Brant or Light-bellied Brent (3)	*Branta bernicla hrota*
Bufflehead	*Bucephala albeola*
Canada Goose (S)	*Branta canadensis*
Aleutian Canada Goose (X)	*Branta canadensis leucopareia*

Common name	Scientific name
Cackling (Canada) Goose (W)	*Branta canadensis minima*
Giant Canada Goose	*Branta canadensis maxima*
Canvasback	*Aythya vallisneria*
Cape Barren Goose	*Cereopsis novaehollandiae*
Carolina or North American Wood Duck	*Aix sponsa*
Chestnut Teal	*Anas castanea*
Comb Duck	*Sarkidiornis melanotos*
African Comb Duck	*Sarkidiornis m. melanotos*
Coscoroba Swan	*Coscoroba coscoroba*
Crested Shelduck	*Tadorna cristata*
Egyptian Goose	*Alopochen aegyptiacus*
Eider Duck	*Somateria mollissima*
Emperor Goose	*Anser canaigicus*
Freckled Duck	*Strictonetta naevosa*
Gadwall	*Anas strepara*
Goldeneye	*Bucephala clangula*
Goosander (American Merganser)	*Mergus merganser*
Greater Scaup	*Aythya marila*
Green-winged Teal	*Anas crecca*
Greylag Goose (J)	*Anser anser*
Harlequin Duck	*Histrionicus histrionicus*
Hawaiian Goose or Nene (Y)	*Branta sanvicensis*
Hooded Merganser	*Mergus cucullatus*
Kelp Goose	*Chloephaga hybrida*
King Eider	*Somateria spectabilis*
Labrador Duck	*Camptorhynchus labradorius*
Laysan Teal	*Anas platyrhynchos laysanensis*
Lesser Scaup	*Aythya affinis*

Common name	Scientific name
Lesser White-fronted Goose	*Anser erythropus*
Long-tailed Duck (Oldsquaw)	*Clangula hyemalis*
Maccoa Duck	*Oxyura maccoa*
Madagascar or Bernier's Teal	*Anas bernieri*
Madagascar Pochard or White-eye	*Aythya innotata*
Magpie Goose	*Anseranas semipalmata*
Mallard	*Anas platyrhynchos*
Mandarin Duck	*Aix galericulata*
Muscovy Duck	*Cairina moschata*
Musk Duck	*Bizura lobata*
Mute Swan	*Cygnus olor*
Pink-eared Duck	*Malacorhynchus membranaceus*
Pink-footed Goose (D)	*Anser brachyrhynchus*
Pink-headed Duck	*Rhodonessa caryophyllacea*
Pintail	*Anas acuta*
Pochard	*Aythya ferina*
Pygmy Goose	*Mettapus sp.*
Red-billed Teal or Red-billed Pintail	*Anas erythrorhynchus*
Red-breasted Goose	*Branta ruficollis*
Red-breasted Merganser	*Mergus serrator*
Redhead	*Aythya americana*
Red Shoveler	*Anas platalea*
Ring-necked Duck	*Aythya collaris*
Ross' Goose	*Anser rossi*
Rosy-billed Pochard	*Netta peposaca*
Ruddy Duck	*Oxyura jamaicensis*
Ruddy-headed Goose	*Chloeophaga rubidiceps*
Salvadori's Duck	*Anas waigiuensis*

Common name	Scientific name
Shelduck	*Tadorna tadorna*
Shoveler	*Anas clypeata*
Snow Goose	*Anser caerulescens*
Lesser Snow (N)	*Anser caerulescens caerulescens*
Greater Snow (O)	*Anser caerulescens atlanticus*
Spur-winged Goose	*Plectopterus gambensis*
Steamer Duck—Flying Steamer Duck	*Trachyeres patachonicus*
Steamer Duck—Flightless Steamer Duck	*Trachyeres pteneres*
Steller's Eider	*Polysticta stelleri*
Swan Goose (A)	*Anser cygnoides*
Teal (Green-winged Teal)	*Anas crecca*
Torrent Duck	*Merganetta armata*
Trumpeter Swan	*Cygnus buccinator*
Tufted Duck	*Aythya fuligula*
Upland or Greater Magellan Goose	*Chloeophaga picta*
Velvet or White-winged Scoter	*Melanitta fusca*
Whistling or Tundra Swan	*Cygnus columbianus columbianus*
White-fronted Goose (E)	*Anser albifrons*
Greenland White-fronted Goose (F)	*Anser albifrons flavirostris*
White-headed Duck	*Oxyura leucocephala*
White-winged Wood Duck	*Cairina scutulata*
Whooper Swan	*Cygnus cygnus*
Wigeon	*Anas penelope*

REFERENCES

Afton, A.D. (1984) Influence of age and time on reproductive performance of female Lesser Scaup. *Auk* **101**, 255–265.

Alerstam, T., Bauer, C-A. and Roos, G. (1974) Spring migration of Eiders *Somateria mollissima* in southern Scandinavia. *Ibis* **116**, 194–210.

Aldrich, J.W. (1973) Disparate sex ratios in waterfowl. pp. 482–489 in Farner, D.S. (ed.), *Breeding biology of birds* Washington, National Academy of Sciences.

Ali, S. (1960) The Pink-headed Duck *Rhodonessa caryophyllacea* (Latham). *Wildfowl Trust Ann. Rep.* **11**, 55–60.

Alison, R.M. (1977) Homing of subadult Oldsquaws. *Auk* **94**, 383–384.

Amat, J.A. (1987*a*) Infertile eggs: a reproductive cost to female dabbling ducks inhabiting unpredictable habitats. *Wildfowl* **38**, 114–116.

Amat, J.A. (1987*b*) Is nest parasitism among ducks advantageous to the host? *American Nat.* **130**, 454–457.

Anderson, D.R. (1975) Population ecology of the Mallard. V. Temporal and geographic estimates of survival. *U.S. Fish and Wildl. Serv. Resource Publ.* **125**.

Anderson, D.R. and Burnham, K.P. (1976) Population ecology of the Mallard. VI. The effect of exploitation on survival. *U.S. Fish and Wildl. Serv. Resource Publ.* **128.**

Anderson, H.G. Food habits of migratory ducks in Illinois. *Ill. Nat. Hist. Surv. Bull.* **27**, 289–344.

Anderson, M.G., Hepp, G.R., McKinney, F. and Owen, M. (1988) Workshop summary: courtship and pairing in winter. pp. 123–131 in Weller, M.W. (ed.), *Waterfowl in Winter* Minneapolis, University of Minnesota Press.

Ankney, C.D. (1977) The use of nutrient reserves by breeding male Lesser Snow Geese (*Chen caerulescens caerulescens*). *Can. J. Zool.* **55**, 1984–1987.

Ankney, C.D. (1979) Does the wing molt cause nutritional stress in Lesser Snow Geese? *Auk* **96**, 68–72.

Ankney, C.D. and Bissett, A.R. (1976) An explanation of egg-weight variation in the Lesser Snow Goose. *J. Wildl. Manage.* **40**, 729–734.

Ankney, C.D. and MacInnes, C.D. (1978) Nutrient reserves and reproductive performance of female Lesser Snow Geese. *Auk* **95**, 459–471.

Anstey, S. (1989) The status and conservation of the white-headed Duck *Oxyura leucocephala*, IWRB Special Publ. 10, 125 pp.

Appleton, T.P. (1982) Rutland Water Nature Reserve: concept, design and management. *Hydrobiologia* **88**, 211–224.

Arnold, T.W., Rohwer, F.C. and Armstrong, T. (1987) Egg viability, nest predation, and adaptive significance of clutch size in prairie ducks. *Amer. Nat.* **130**, 643–653.

Ashcroft, R.E. (1976) A function of the pairbond in the Common Eider. *Wildfowl* **27**, 101–105.

Bacon, P.E. (1980) Status and dynamics of a Mute Swan population near Oxford between 1976 and 1978. *Wildfowl* **31**, 37–50.

Baillie, S.R. and Milne, H. (1982) The influence of female age on breeding in the Eider *Somateria mollissima*. *Bird Study* **29**, 55–66.

Baker, R.R. (1978) *The Evolutionary Ecology of Animal Migrations* London, Hodder and Stoughton.

Baldessarre, G.A., Quinlan, E.E. and Bolen, E.G. (1988) Mobility and site fidelity of Green-winged Teal wintering on the southern high plains of Texas. pp. 483–493 in Weller, M.W. (ed.), *Waterfowl in Winter* Minneapolis, University of Minnesota Press.

Ball, I.J., Gilmer, D.S., Cowardin, L.M. and Riechmann, J.H. (1975) Survival of Wood Duck and Mallard broods in north-central Minnesota. *J. Wildl. Manage.* **39**, 776–780.

Ball, I.J., Frost, P.G.H., Siegfried, W.R. and McKinney, F. (1978) Territories and local movements of African Black Ducks. *Wildfowl* **29**, 61–79.

Banko, W.E. (1960) The Trumpeter Swan: its history, habits, and population in the United States. *North American Fauna* **63**, Washington, Bureau of Sport, Fisheries and Wildlife.

Barnes, G.G. and Thomas, V.G. (1986) Digestive organ morphology, diet and guild structure of North American Anatidae. *Canadian Journal of Zoology* **65**, 1812–1817.

Batten, L.A. and Swift, J.A. (1981) British criteria for calling a ban on wildfowling in severe weather. pp. 181–189 in Scott, D.A. and Smart, M. (eds), *Proc. 2nd Tech. Meeting on Western Palearctic Migratory Bird Management* Paris 1979. Slimbridge, IWRB.

Bedard, J., Nadeau, A. and Gauthier, G. (1986) Effects of spring grazing by Greater Snow Geese on hay production. *J. Appl. Ecol.* **23**, 65–75.

Bellrose, F.C. (1959) Lead poisoning as a mortality factor in waterfowl populations. *Ill. Nat. Hist. Surv. Bull.* **27**, 235–288.

Bellrose, F.C. (1968) Waterfowl migration corridors east of the Rocky Mountains in the United States. *Ill. Nat. Hist. Surv. Biol. Notes* **61**, 23pp.

Bellrose, F.C. (1978) *Ducks, Geese and Swans of North America.* Second Edition. Harrisburg, Stackpole Books.

Bengtson, S-A. (1972) Reproduction and fluctuations in the size of the duck populations at Lake Myvatn, Iceland. *Oikos* **23**, 35–58.

Bennett, J.W. and Bolen, E.G. (1978) Stress response in wintering Green-winged Teal. *J. Wildl. Manage.* **42**, 81–86.

Bethke, R.W. and Thomas, V.G. (1988) Differences in flight and heart muscle mass among geese, dabbling ducks and diving ducks relative to habitat use. *Canadian Journal of Zoology* **66**, 2024–2028.

Birkhead, M. and Perrins, C. (1986) *The Mute Swan.* London, Croom Helm.

Birkhead, M.E., Bacon, P.J. and Walter, P. (1983) Factors affecting the breeding success of the Mute Swan. *J. Anim. Ecol.* **52**, 727–741.

Black, J.M. (1988) Preflight signalling in swans: a mechanism for group cohesion and flock formation. *Ethology* **79**, 143–157.

Black, J.M. and Owen, M. (1987) Determinant factors of social rank in goose flocks: acquisition of social rank in young geese. *Behaviour* **102**, 129–146.

Black, J.M. and Owen, M. (1988) Variations in pair bond and agonistic behaviors in Barnacle Geese on the wintering grounds. pp. 39–57 in Weller, M.W. (ed.) *Waterfowl in Winter* Minneapolis, University of Minnesota Press.

Black, J.M. and Owen, M. (1989*a*) Parent–offspring relationships in wintering Barnacle Geese. *Anim. Behav.* **37**, 187–198.

Black, J.M. and Owen, M. (1989*b*) Agonistic behaviour in Barnacle Goose flocks: assessment, investment and reproductive success. *Anim. Behav.* **37**, 199–209.

Black, J.M. and Rees, E.C. (1984) The structure and behaviour of the Whooper Swan population wintering at Caerlaverock, Dumfries and Galloway, Scotland: an introductory study. *Wildfowl* **35**, 21–36.

Black, J.M., Deerenberg, C. and Owen, M. (1990) Foraging behaviour and site selection of Barnacle Geese in a traditional and newly established spring staging habitat. *Ardea* **78**, in press.

Blokpoel, H. (1974) Migration of Lesser Snow and Blue Geese in spring across southern Manitoba. Part I. Distribution, chronology, directions, numbers, heights and speeds. *Can. Wildl. Serv. Rep. Ser.* **28**, 30pp.

Blokpoel, H. and Gauthier, M.C. (1975) Migration of Lesser Snow and Blue Geese in spring across southern Manitoba. Part 2. Influene of weather and prediction of major flights. *Can. Wildl. Serv. Rep. Ser.* **32**, 30pp.

Bousfield, M.A. and Syroechkovskiy, Ye.V. (1985) A review of Soviet research on the Lesser Snow Goose on Wrangel Island. *Wildfowl* **36**, 13–20.

Boyd, H. (1953) On encounters between wild White-fronted Geese in winter flocks. *Behaviour* **5**, 85–129.

Boyd, H. (1964*a*) Wild geese at the New grounds, 1962–63. *Wildfowl Trust Ann. Rep.* **15**, 19.

Boyd, H. (1964*b*) Wildfowl and other water birds found dead in England and Wales in January–March 1963. *Waterfowl Trust Ann. Rep.* **15**, 20–22.

Boyd, H. (1976) Estimates of total numbers in the Hudson Bay population of Lesser Snow Geese 1964–1973. *Can. Wildl. Serv. Prog. Notes* **63**.

Boyd, H. (1980) Waterfowl crop damage prevention and compensation programs in the Canadian Prairie Provinces. pp. 20–27 in Wright, E.N., Inglis, I.R. and Feare, C.J. (eds.), *Bird Problems in Agriculture* Croydon, BPCC Publications.

Boyd, H. (1983) Intensive regulation of duck hunting in North America: its purpose and achievements. Occasional Paper 50, Canadian Wildlife Service. Ottawa, CWS.

Boyd, H. and Ogilvie, M.A. (1969) Changes in the British-wintering population of the Pink-footed Goose from 1950 to 1975. *Wildfowl* **20**, 33–46.

Braithwaite, L.W. (1976) Breeding seasons of waterfowl in Australia. *Proc. 16th Int. Ornithol. Congr.*, 235–247.

Braithwaite, L.W. (1982) Ecological studies of the Black Swan. IV. The timing and success of breeding on two nearby lakes on the southern tablelands of New South Wales. *Australian Wildlife Research* **91**, 261–275.

Brodsky, L.M. and Weatherhead, P.J. (1985) Time and energy constraints on courtship in wintering American Black Ducks. *Condor* **87**, 33–36.

Brown, J.L. (1987) Helping and communal breeding in birds: ecology and evolution. Princeton, N.J., Princeton University Press.

Bruinderink, G.W.T. (1987) *Wilde ganzen en culturgrasland in Nederland.* Unpubl. Thesis, Landbouwuniversiteit te Wageningen.

Bruzual, J.J. and Bruzual, I.B. (1983) Feeding habits of whistling ducks in the Calabozo ricefields, Venezuela, during the non-reproductive period. *Wildfowl* **34**, 20–26.

Bryant, D.M. and Leng, J. (1975) Feeding distribution and behaviour of Shelduck in relation to food supply. *Wildfowl* **26**, 20–30.

Buchsbaum, Valela, I. and Swain, T. (1984) The role of phenolic compounds and other plant constituents in feeding by Canada Geese in a coastal marsh. *Oecologia* **63**, 343–349.

Buxton, N.E. (1975) *The feeding behaviour and food supply of the Common Shelduck* (Tadorna tadorna) *on the Ythan Estuary, Aberdeenshire.* Unpubl. PhD Thesis, University of Aberdeen.

Cabot, D., Nairn, R., Newton, S. and Viney, M. (1984) *Biological expedition to Jameson Land, Greenland 1984.* Dublin, Barnacle Books.

Caldwell, P.J. and Cornwell, G.W. (1975) Incubation behaviour and temperatures of the Mallard duck. *Auk* **92**, 706–731.

Campbell, C.R.G. and Ogilvie, M.A. (1982) Failure of Whooper Swan to moult wing feathers. *Brit. Birds* **75**, 578.

Campredon, P. (1981) Hivernage du canaerd siffleur *Anas penelope* L. en Camargue (France). Stationnements et activities. *Alauda* **49**, 161–193.

Cargill, S.M. and Jeffries, R.L. (1986) The effects of grazing by Lesser Snow Geese on the vegetation of a sub-arctic saltmarsh. *J. Appl. Ecol.* **21**, 669–686.

Charman, K. (1980) Feeding ecology and energetics of the Dark-bellied Brent Goose *Branta bernicla bernicla* in Essex and Kent. pp. 451–465 in Jeffries, R.L. and Davy, A.J. (eds), *Ecological Processes in Coastal Environments* Oxford, Blackwell Scientific Publications.

Clawson, R.L., Hartman, G.W. and Fredrickson, L.H. (1979) Dump nesting in a Missouri Wood Duck population. *J. Wildl. Manage.* **43**, 347–355.

Clutton-Brook, T.H. (ed.) (1988) *Reproductive Success.* Chicago, Chicago University Press.

Cole, R.W. (1979) The relationship between weight at hatch and survival and growth of wild Lesser Snow Geese. M.S. Thesis, University of Western Ontario, London.

Coleman, A.E. and Minton, C.D.T. (1979) Pairing and breeding of Mute Swans in relation to natal area. *Wildfowl* **30**, 27–30.

Coleman, A.E. and Minton, C.D.T. (1980) Mortality of Mute Swan progeny in an area of south Staffordshire. *Wildfowl* **31** 22–28.

Collar, N.J. and Andrew, P. (1988) *Birds to Watch.* Cambridge, ICBP.

Cooch, E.G., Lank, D.B., Rockwell, R.F. and Cooke, F. (1989) Long-term decline in fecundity in a Snow Goose population: evidence of density dependence? *J. Anim. Ecol.* **58**, 711–726.

Cook, A.S. (1973) Shell thinning in avian eggs by environmental pollutants. *Environmental Pollution* **4**, 85–152.

Cooke, F. (1990) Density related aspects of reproduction in the Lesser Snow Goose at La Perouse Bay, Manitoba. Perrins, C.M. (ed.), *Proc. Symp on Population Dynamics and their relevance to Conservation.* The Camargue, France, December 1988.

Cooke, F., MacInnes, C.D. and Prevett, J.P. (1975) Gene flow between breeding populations of the Lesser Snow Goose. *Auk* **92**, 493–510.

Craighead, J.J. and Stockstad, D.S. (1964) Breeding age of Canada Geese. *J. Wildl. Manage.* **28**, 57–64.

Danell, K. and Sjoberg, K. (1977) Seasonal emergence of Chironomids in relation to egglaying and hatching of ducks on a restored lake (northern Sweden). *Wildfowl* **28**, 129–135.

Davies, J.C. and Cooke, F. (1983) Annual nesting productivity in Snow Geese: prairie droughts and Arctic springs. *J. Wildl. Manage.* **47**, 271–280.

Delacour, J. (1954–1964) *The Waterfowl of the World* Vols 1–4. London, Country Life.

Derrickson, S.R. (1978) The mobility of breeding Pintails. *Auk* **95**, 104–114.

Dorward, D.F., Norman, F.I. and Cowling, S.J. (1980) The Cape Barren Goose in Victoria, Australia: management related to agriculture. *Wildfowl* **31**, 144–150.

Douthwaite, R.J. (1976) Weight changes and wing moult in the Red-billed Teal. *Wildfowl* **27**, 123–127.

Dow, H. and Fredga, S. (1983) Breeding and natal dispersal of the Goldeneye *Bucephala clangula. J. Anim. Ecol.* **52**, 681–695.

Draulans, D. (1982) Foraging and size selection of mussels by the Tufted Duck, *Aythya fuligula. J. Anim. Ecol.* **51**, 943–956.

Drent, R.H. and Daan, S. (1980) The prudent parent: energetic adjustments in avian breeding. *Ardea* **68**, 225–252.

Drent, R.H. and Swierstra, P. (1977) Goose flocks and food finding: field experiments with Barnacle Geese in winter. *Wildfowl* **28**, 15–20.

Drent, R.H., Weijand, B. and Ebbinge, B. (1980) Balancing the energy budgets of arctic breeding geese throughout the annual cycle: a progress report. *Verh. Orn. Ges. Bayern.* **23**, 239–264.

Dnebbert, H.F. and Lokemoon, J.T. (1980) High duck nesting success in a predator-reduced environment. *J. Wildl. Manage.* **44**, 428–437.

Dzubin, A. (1969) Assessing breeding populations of ducks by ground counts. *Can. Wildl. Serv. Report Series* No. 6. Ottawa, CWS.

Dzubin, A. and Gollop, J.B. (1972) Aspects of Mallard breeding ecology in the Canadian parkland and grassland. pp. 113–152 in *Population Ecology of Migratory Birds. U.S. Bureau of Sport Fish. and Wildl. Res.* **2**.

Dzubin, A., Boyd, H. and Stephen, W.J. (1975) Blue and Snow Goose distribution in the Mississippi and Central Flyways. *Can. Wildl. Serv. Prelim. Rep.* **1**, 81pp.

Ebbinge, B. (1989) A multifactorial explanation for variation in breeding performance of Brent Geese *Branta bernicla. Ibis* **131**, 196–204.

Ebbinge, B. (1990) The effect of shooting on goose populations. *Ardea* **78**, in press.

Ebbinge, B. and Ebbinge-Dallmeijer, D. (1977) Barnacle Geese (*Branta leucopsis*) in the Arctic summer—a reconnaisance trip to Svalbard. *Norsk Polarinstitutt Aarbok 1975*, 119–138.

Einarsson, A. (1987) Distributional movements of Barrow's Goldeneye *Bucephala islandica* young in relation to food. *Ibis* **130**, 153–163.

Ekman, S. (1922) *Djurvaldens utbredningshistoria paa Scandinaviska havlon*. Stockholm.
Eldridge, J.L. (1986*a*) Observations on a pair of Torrent Ducks. *Wildfowl* **37**, 113–122.
Eldridge, J.L. (1986*b*) Territoriality in a river specialist: the Blue Duck. *Wildfowl* **37**, 123–135.
Elkins, N. (1979) High altitude flight by swans. *Brit. Birds* **72**, 238–239.
Eltringham, S.K. (1974) The survival of broods of the Egyptian Goose in Uganda. *Wildfowl* **25**, 41–48.
Eriksson, M.O.G. (1979) Competition between freshwater fish and Goldeneye *Bucephala clangula* (L.) for common prey. *Oecologia* (*Berl.*) **41**, 99–107.
Erskine, A.J. (1961) Nest-site tenacity and homing in the Bufflehead. *Auk* **78**, 389–396.
Erskine, A.J. (1972) *Buffleheads*. Ottawa Canadian Wildlife Service, Monograph Series 4.
Evans, M.E. (1979) Aspects of the life cycle of the Bewick's Swan, based on recognition of individuals at a wintering site. *Bird Study* **26**, 149–162.
Evans, M.E. and Kear, J. (1978) Weights and measurements of Bewick's Swans during winter. *Wildfowl* **29**, 118–122.
Evarts, S. and Williams, C.J. (1987) Multiple parernity in a wild population of Mallards. *Auk* **104**, 597–602.
Finney, G. and Cooke, F. (1978) Reproductive habits in the Snow Goose: the influence of female age. *Condor* **80**, 147–158.
Findlay, C.S. and Cooke, F. (1982) Synchrony in the Lesser Snow Goose *Anser caerulescens caerulescens*. II. The adaptive value of reproductive synchrony. *Evolution* **36**, 786–799.
Flickinger, E.L. and Bolen, E.G. (1979) Weights of Lesser Snow Geese taken on their winter range. *J. Wildl. Manage.* **43**, 531–533.
Fox, A.D. and Mitchell, C.R. (1988) Migration and seasonal distribution of Gadwall breeding in Britain and Ireland: a preliminary assessment. *Wildfowl* **39**, 145–152.
Fox, A.D., Owen, M., Salmon, D.G. and Ogilvie, M.A. (1989) Population dynamics of Icelandic-nesting geese, 1960–1987. *Ornis Scand.* **20**, 289–297.
Fretwell, S.D. (1972) *Populations in a Seasonal Environment*. Princeton, Princeton University Press.
Friend, M. and Pearson, G.L. (1973) Duck Plague (Duck Virus Entritis) in wild Waterfowl. *Rep. U.S. Bur. Sport Fisheries and Wildl.* 16pp.
Frith, H.J. (1962) Movements of the Grey Teal *Anas gibberifrons* Muller (Anatidae). *CSIRO Wildl. Res.* **7**, 50–70.
Frith, H.J. (1982) *Waterfowl in Australia*. Revised Edition. Sydney, Angus & Robertson Publishers.
Frith, H.J. and Davies, S.J. (1961) Ecology of the Magpie Goose (*Anseranas semipalmata*) Latham (Anatidae). *CSIRO Wildl. Res.* **6**, 91–141.
Gauthier, G. (1988) Territorial behaviour, forced copulations and mixed reproductive strategy in ducks. *Wildfowl* **39**, 102–114.
Gauthier, G. and Smith, J.N.M. (1987) Territorial behaviour, nest site availability, and breeding density in Buffleheads. *J. Anim. Ecol.* **56**, 171–184.
Gauthreaux, S.A. (1982) The ecology and evolution of avian migration systems. pp. 93–168 in Farner, D.S., King, J.R. and Parkes, K.C. (eds), *Avian Biology* New York, Academic Press.
Gauvin, J. and Reed, A. (1987) A simulation model for the Greater Snow Goose population. *Can. Wildl. Serv. Occ. Paper* **64**.
Giroux, J.-F. (1981) Use of artificial islands by nesting waterfowl in southeastern Alberta. *J. Wildl. Manage.* **45**, 669–679.
Goodburn, B.F. (1984) Mate guarding in the Mallard *Anas platyrhynchos*. *Ornis Scand.* **15**, 261–265.
Goode, D.A. (1981) *Lead poisoning in swans*. Rep. Nature Conservancy Council Working Group. London, NCC.
Gorman, M.L. and Milne, H. (1972) Creche behaviour in the Common Eider *Somateria m. mollissima. L. Ornis Scand.* **3**, 21–25.
Goudie, R.I. and Ankney, C.D. (1986) Body size, activity budgets, and diets of sea ducks wintering in Newfoundland. *Ecology* **67**, 1475–1482.

Gullestad, N., Owen, M. and Nugent, M.J. (1984) Numbers and distribution of Barnacle Geese *Branta leucopsis* on Norwegian staging islands and the importance of the staging area to the Svalbard population. *Norsk Polarinstitiutt Skrifter* **181**, 57–65.
Hamann, J., Andrews, B. and Cooke, F. (1986) The role of follicular atresia in inter- and intra-seasonal clutch size variation in Lesser Snow Geese (*Anser caerulescens caerulescens*). *J. Anim. Ecol.* **55**, 481–489.
Hamilton, W.D. (1971) Geometry for the selfish herd. *J. Theor. Biol.* **31**, 295–311.
Hankla, D.J. (1968) Summary of Canada Goose transplant program on nine national wildlife refuges in the southeast. pp. 105–111 in Hine R.I. and Schoenfeld, C. (eds), *Canada Goose Management* Madison, Dembar Educational Inc.
Hansen, H.A. and McKnight, D.E. (1964) Emigration of drought-displaced ducks to the Arctic. *Trans. N. Amer. Wildl. Nat. Resources Conf.* **29**, 119–126.
Hanson, H.C. (1962) The dynamics of condition factors in Canada Geese and their relation to seasonal stresses. *Arctic Inst. of N. Amer. Tech. Paper* No. 12, 68pp.
Harradine, J. (1982) Some mortality patterns of Greater Magellan Geese in the Falkland Islands. *Wildfowl* **33**, 7–11.
Harrison, J.G. and Wardell, J. (1970) WAGBI duck to supplement wild populations. pp. 195–209 in Sedgwick, N.M., Whitaker, P. and Harrison, J.G. (eds), *The New Wildfowler in the 1970s* London, Barrie and Jenkins.
Harwood, J. (1977) Summer feeding ecology of Lesser Snow Geese. *J. Wildl. Manage.* **41**, 48–55.
Hawkins, L. (1986) Nesting behaviour of male and female Whistling Swans and implications of male incubation. *Wildfowl* **37**, 5–27.
Heitmeyer, M.E. and Fredrikson, L.H. (1981) Do wetland conditions in the Mississippi Delta hardwoods influence Mallard recruitment? *Trans. N. Amer. Wildl. Nat. Resources Conf.* **46**, 44–57.
Henny, C.J. and Holgerson, N.E. (1974) Range expansion and population increase in the Gadwall in eastern North America. *Wildfowl* **25**, 95–101.
Hepp, G.R. and Hair, J.D. (1983) Reproductive behaviour and pairing chronology in wintering dabbling ducks. *Wilson Bull.* **95**, 675–682.
Heusmann, H.W. (1972) Survival of Wood Duck broods from dump nests. *J. Wildl. Manage.* **36**, 620–624.
Hill, D.A. (1984) Population regulation in the Mallard (*Anas platyrhynchos*). *J. Anim. Ecol.* **53**, 191–202.
Hill, D.A., Wright, R. and Street, M. (1987) Survival of Mallard ducklings *Anas platyrhynchos* and competition with fish for invertebrates on a flooded gravel quarry in England. *Ibis* **129**, 159–167.
Hochbaum, H.A. (1944) *The Canvasback on a prairie marsh.* Washington D.C., Am. Wildl. Inst.
Hochbaum. H.A. (1955) *Travels and Traditions of Waterfowl.* Minneapolis, University of Minnesota Press.
Hohner, W.L., Taylor, T.S. and Weller, M.W. (1988) Annual body weight change in Ring-necked Ducks (*Aythya collaris*). pp. 257–269 in Weller, M.W. (ed.), *Waterfowl in Winter* Minneapolis, University of Minnesota Press.
Hori, J. (1987) Distribution, dispersion and regulation in a population of the Common Shelduck. *Wildfowl* **38**, 127–142.
Imber, M.J. (1968) Sex ratios in Canada Goose populations. *J. Wildl. Manage.* **32**, 905–920.
Inglis, I.J. and Lazarus, J. (1981) Vigilance and flock size in Brent Geese: the edge effect. *Z. Tierpsychol.* **57**, 193–200.
Joensen, A.H. (1973) Moult migration and wing feather moult of seaducks in Denmark. *Dan. Rev. of Game Biol. 8* No. 4, 42pp.
Joensen, A.H. (1976) Oil pollution and seaducks in Denmark. pp. 15–17 in Anderson, A. and Fredga, S. (eds), *Proc. Symp. on Sea Ducks, Stockholm 1975.* Slimbridge, IWRB.

Johnsgard, P.A. (1965) *Handbook of Waterfowl Behaviour*. Ithaca, New York, Cornell University Press.

Johnsgard, P.A. (1973) Proximate and ultimate determinants of clutch size in Anatidae. *Wildfowl* **24**, 144–149.

Johnson, D.H. and Grier, J.W. (1988) Determinants of breeding distributions of ducks. *Wildl. Monogr.* **100**, 1–37.

Johnson, J.C. and Raveling, D.G. (1988) Weak family associations in Cackling Geese during winter: effects of body size and food resources on goose social organization. pp. 257–269 in Weller, M.W. (ed.), *Waterfowl in Winter* Minneapolis, University of Minnesota Press.

Jones, R.D. and Jones, D.M. (1966) The progress of family disintegration in the Black Brant. *Wildfowl Trust Ann. Rep.* **17**, 75–78.

Jorde, D.G. and Owen, R.B. (1988) The need for nocturnal activity and energy budgets of waterfowl. pp. 169–180 in Weller, M.W. (ed.), *Waterfowl in Winter* Minneapolis, University of Minnesota Press.

Jorde, D.G., Krapu, G.L., Crawford, R.D. and Hay, M.A. (1984) Effects of weather on habitat selection and behaviour of Mallards wintering in Nebraska. *Condor* **86**, 258–265.

Kadlec, J.A. (1962) Effects of a drawdown on a waterfowl impoundment. *Ecology* **43**, 267–281.

Kear, J. (1965) The internal food reserves of hatching Mallard ducklings. *J. Wildl. Manage.* **29**, 523–528.

Kear, J. (1967) Experiments with young nidifugous birds on a visual cliff. *Wildfowl Trust Ann. Rep.* **18**, 122–124.

Kear, J. (1970) The adaptive radiation of parental care in waterfowl. pp. 357–392 in Crook, J.H. (ed.), *Social Behaviour in Birds and Mammals* London, Academic Press.

Kear, J. (1975) Salvadori's Duck of New Guinea. *Wildfowl* **26**, 104–111.

Kear, J. and Berger, A.J. (1980) *The Hawaiian Goose*. Calton, T & A.D. Poyser.

Kear, J. and Scarlett, R.J. (1970) The Auckland Islands Merganser. *Wildfowl* **21**, 78–86.28.

Kear, J. and Steel T.H. (1971) Aspects of social behaviour in the Blue Duck. *Notornis* **18**, 187–198.

Kear, J. and Williams, G. (1978) Waterfowl at risk. *Wildfowl* **29**, 5–21.

Kehoe, F.P. and Thomas, V.G. (1986) A comparison of interspecific differences in the morphology of external and internal feeding apparatus among North American Anatidae. *Canadian Journal of Zoology* **65**, 1818–1822.

Kerbes, R.H. (1975) The nesting population of Lesser Snow Geese in the eastern Canadian Arctic: a photographic inventory of June 1973. *Can. Wildl. Service Rep. Series* **35**, 47pp.

King, J.G. (1973) A cosmopolitan duck moulting resort: Takslesluk Lake Alaska. *Wildfowl* **24**, 103–109.

Kirby, R.E. and Ferrigno, F. (1980) Winter, waterfowl and the salt marsh. *New Jersey Outdoors* **7**, 10–13.

Korschgen, C.E. (1977), Breeding stress of female Eiders in Maine. *J. Wildl. Manage.* **41**, 360–373.

Korschgen, C.E., George, L.S., and Green, W.L. (1988) Feeding ecology of Canvasbacks staging on pool 7 of the Upper Mississippi River. pp. 237–249 in Weller, M.W. (ed.), *Waterfowl in Winter* Minneapolis, University of Minnesota Press.

Krapu, G.L. (1974) Feeding ecology of pintail hens during reproduction. *Auk* **91**, 278–290.

Krapu, G.L. (1981) The role of nutrient reserves in Mallard reproduction. *Auk* **98**, 29–38.

Krapu, G.L. and Doty, H.A. (1979) Age-related aspects of Mallard reproduction. *Wildfowl* **30**, 35–39.

Krebs, J.R. and Davies, N.B. (1987) *An Introduction to Behavioural Ecology*. Second Edition. Oxford, Blackwell Scientific Publications.

Krivonosov, G. (1970) The Volga Delta as a wildfowl haunt. pp. 70–72 in Isakov, Y.A. (ed.), *Proc. Int. Regional Meeting on Conserv. of Wildfowl Resources. Leningrad 1968*. Moscow.

Lack, D. (1954) *The Natural Regulation of Animal Numbers*. Oxford, Clarendon Press.

Lack, D. (1967) The significance of clutch-size in waterfowl. *Wildfowl* **18**, 125–128.
Lack, D. (1968) The proportion of yolk in the eggs of waterfowl. *Wildfowl* **19**, 67–69.
Lamprecht, J. (1989) Mate guarding in geese: awaiting female receptivity, protection of paternity or support of female feeding? pp. 48–66 in Rasa, A.E., Vogel, C. and Voland, E. (eds), *The Sociobiology of Sexual and Reproductive Strategies*. London, Chapman and Hall.
Lamprecht, J. and Burhow, H. (1987) Harem polygyny in Bar-headed Geese (*Anser indicus*). *Ardea* **75**, 285–292.
Lank, D.B., Cooch, E.G., Rockwell, R.F. and Cooke, F. (1989) Environmental and demographic correlates of intraspecific nest parasitism in Lesser Snow Geese *Chen caerulescens caerulescens*. *J. Anim. Ecol.* **58**, 29–45.
Larsson, K., Forslund, P., Gustafsson, L. and Ebbinge, B.S. (1988) From the high Arctic to the Baltic: the successful establishment of a Barnacle Goose *Branta leucopsis* population on Gotland, Sweden. *Ornis Scand.* **19**, 182–189.
Lazarus, J. (1978) Vigilance, flock size and domain of danger size in the White-fronted Goose. *Wildfowl* **29**, 135–145.
Lazarus, J. and Inglis, I.R. (1978) The breeding behaviour of the Pink-footed Goose: parental care and vigilant behaviour during the fledging period. *Behaviour* **65**, 62–88.
Lebret, T. (1950) The sex ratios and the proportion of adult drakes of Teal, Pintail, Shoveler and Wigeon in the Netherlands based on field counts made during autumn, winter and spring. *Ardea* **38**, 1–18.
Lee, F.B. and Kruse, A.D. (1973) High survival and homing rate of hand-reared wild-strain Mallards. *J. Wildl. Manage.* **37**, 154–159.
Lensink, C.J. (1973) Population structure and productivity of Whistling Swans on the Yukon Delta, Alaska. *Wildfowl* **24**, 21–25.
Lessells, C.M. (1985) Natal and breeding dispersal of Canada Geese *Branta canadensis*. *Ibis* **127**, 31–41.
Lessells, C.M. (1986) Brood size in Canada Geese: a manipulation experiment. *J. Anim. Ecol.* **55**, 669–689.
Linduska, J.P. (1972) Waterfowl utilization in the United States. pp. 137–148 in Carp, E. (ed.), *Proc. Int. Conf. on the Conservation of Wetlands and Waterfowl, Ramsar, Iran, 1971*. Slimbridge, IWRB.
Linduska, J.P. (1982) Sanctuaries in waterfowl management in the USA. pp. 319–330 in Scott, D.A. (ed.), *Managing Wetlands and their Birds*. Slimbridge, IWRB.
Lorenz, K.Z. (1941) Vergleichende bewegungsstudien an Anatinen. *J. Ornithol.* **89** (Suppl.), 194–294.
MacInnes, C.D. and Misra, R.K. (1972) Predation on Canada Goose nests at McConnell River, Northwest Territories. *J. Wildl. Manage.* **36**, 414–422.
MacInnes, C.D., Davis, R.A., Jones, R.N., Lieff, B. and Pakulak, A.J. (1974) Reproductive efficiency of McConnell River small Canada Geese. *J. Wildl. Manage.* **38**, 686–707.
Madsen, J. (1985) Habitat selection of farmland feeding geese in west Jutland, Denmark: an example of a niche shift. *Ornis Scand.* **16**, 40–144.
Madsen, J. and Mortensen, C.E. (1987) Habitat exploitation and interspecific competition of moulting geese in east Greenland. *Ibis* **129**, 25–44.
Martin, K., Cooch, F.G., Rockwell, R.F. and Cooke, F. (1985) Reproductive performance in Lesser Snow Geese: are two parents essential? *Behav. Ecol. Sociobiol.* **17**, 257–263.
Mathiasson, S. (1973) Moulting grounds of Mute Swans (*Cygnus olor*) in Sweden, their origin and relation to the population dynamics of Mute Swans in the Baltic area. *Viltrevy* **8**, 399–452.
Mathiasson, S. (1974) A moulting population of non-breeding Mute Swans with special reference to flight-feather moult, feeding ecology and habitat selection. *Wildfowl* **24**, 43–43.
Matthews, G.V.T. (1981) The conservation of migratory birds. pp. 231–239 in Aidley, R. (ed.), *Animal Migration*. Cambridge, University Press.

Mayhew, P. and Houston, D. (1989) Feeding site selection by Wigeon *Anas penelope* in relation to water. *Ibis* **131**, 1–8.

McKinney, F. (1965) Spacing and chasing in breeding ducks. *Wildfowl Trust Ann. Rep.* **16**, 92–106.

McKinney, F. (1986) Ecological factors influencing the social systems of migratory dabbling ducks. pp. 153–179 in Rubenstein, D.I. and Wrangham, R.W. (eds) *Ecological aspects of social evolution* Princeton, N.J., University of Princeton Press.

McKinney, F., Siegfried, W.R., Ball, I.J. and Frost, P.G.H. (1978) Behavioural specialisations for river life in the African Black Duck (*Anas sparsa* Eyton). *Z. Tierpsychol.* **48**, 349–400.

McKinney, F., Derrickson, S.R. and Mineau, P. (1983) Forced copulation in waterfowl. *Behaviour* **86**, 250–294.

Meltofte, H. (1978) A breeding association between Eiders and tethered huskies in north-east Greenland. *Wildfowl* **29**, 45–54.

Meltofte, H. (1978) Skudeffektivitet ved intensiv kystfuglejagt i Danmark. En pilotundersøgelse *Dansk Orn. Foren. Tiddskr.* **72**, 217–221.

Mendenhall, V. (1976) Survival and causes of mortality in Eider ducklings on the Ythan Estuary, Aberdeenshire. *Wildfowl* **27**, 160.

Mickelson, P.G. (1975) Breeding biology of Cackling Geese and associated species on the Yukon-Kuskokwim Delta, Alaska. *Wildl. Monogr.* **45**, 35pp.

Mikula, E.J., Martz, G.F. and Bennett, C.L. (1972) Field evaluation of three types of waterfowl hunting regulations. *J. Wildl. Manage.* **36**, 441–459.

Miller, K.J. (1976) Activity patterns, vocalisations and site selection in nesting Blue-winged Teal. *Wildfowl* **27**, 33–43.

Milne, H. (1976) Body weights and carcass composition of the Common Eider. *Wildfowl* **27**, 115–122.

Minton, C.D.T. (1968) Pairing and breeding in Mute Swans. *Wildfowl* **19**, 41–60.

Munro, J. and Bedard, J. (1977) Gull predation and creching behaviour in the Common Eider. *J. Anim. Ecol.* **46**, 799–810.

Murton, R.K. and Kear, J. (1973) The nature and evolution of the photoperiodic control of reproduction in wildfowl of the family Anatidae. *J. Reprod. Fert. Suppl.* **19**, 67–84.

Neuchterlein, G.L. and Storer, R.W. (1985) Aggressive behaviour and interspecific killing by Flying Steamer Ducks in Argentina. *Condor* **87**, 87–91.

Newton, I and Campbell, C.R.G. (1973) Feeding of geese on farmland in east-central Scotland. *Journal of Applied Ecology* **10**, 781–801.

Newton, I. and Campbell, C.R.G. (1975) Breeding of ducks at Loch Leven, Kinross. *Wildfowl* **26**, 83–103.

Newton, I., Thom, V.M. and Brotherston, W. (1973) Behaviour and distribution of wild geese in south-east Scotland. *Wildfowl* **24**, 111–121.

Nicholls, J.D. and Haramis, G.M. (1980) Sex specific differences in winter distribution patterns of Canvasbacks. *Condor* **82**, 406–416.

Nicholls, J.D. and Hines, J.E. (1983) The relationship between harvest and survival rates of mallards: A straightforward approach with partioned data sets. *J. Wildl. Manage.* **47**, 334–348.

Nicholls, J.D., Conroy, M.J., Anderson, D.R. and Burnham, K.P. (1984) Compensatory mortality in waterfowl populations: A review of the evidence and implications for research and management. *N. Amer. Wildl. and Nat. Resources Conf.* **49**, 535–553.

Nilsson, L. (1970) Food-seeking activity of south Swedish diving ducks in the non-breeding season. *Oikos* **21**, 145–154.

Nilsson, L. (1972) Habitat selection, food choice and feeding habits of diving ducks in coastal waters of south Sweden during the non-breeding season. *Ornis Scandinavica* **3**, 55–78.

Nilsson, L. (1979) Variation in the production of young swans wintering in Sweden. *Wildfowl* **30**, 129–134.

Nilsson, L. (1984) The impact of hard winters on waterfowl populations of southern Sweden. *Wildfowl* **35**, 71–80.

Norman, F.I. and McKinney, F. (1987) Clutches, broods, and brood care behaviour in the Chestnut Teal. *Wildfowl* **38**, 117–126.

Nudds, T.B. (1978) Comments on Calverley and Boag's (1977) hypothesis on displaced ducks and an evolutionary alternative. *Can. J. Zool.* **56**, 2239–2241.

O'Briain, M. (1987) Families and other social groups of Brent Geese in winter. Unpubl. Rep. University of Dublin.

Ogilvie, M.A. (1967) Population changes and mortality of Mute Swans in Britain. *Wildfowl* **18**, 64–73.

Ogilvie, M.A. (1981) Hard weather movements of *Anas crecca* ringed in western Europe—a preliminary computer analysis. *Proc IWRB Symp. Alushta, USSR* 119–135.

Olney, P.J.S. (1963) The food and feeding habits of Teal *Anas crecca* L. *Proc. Zool. Soc. Lond.* **140**, 169–210.

Olney, P.J.S. (1965) The autumn and winter feeding of certain sympatric ducks. *Trans. VI Congress Int. Union of Game Biologists Bournemouth*, 309–322.

Oring, L.W. (1982) Avian Mating systems. pp. 1–92 in Farner, D.S., King, J.R. and Parkes, K.C. (eds), *Avian Biology Vol. VI.* New York, Academic Press.

Owen, M. (1972*a*) Some factors affecting food intake and selection in White-fronted Geese. *J. Anim. Ecol.* **41**, 79–92.

Owen, M. (1972*b*) Movements and ecology of White-fronted Geese at the New Grounds, Slimbridge. *J. Appl. Ecol.* **9**, 385–398.

Owen, M. (1973) The management of grassland areas for wintering geese. *Wildfowl* **24**, 123–130.

Owen, M. (1975) Cutting and fertilizing grassland for winter goose management. *J. Wildl. Manage.* **39**, 163–167.

Owen, M. (1976) Factors affecting the distribution of geese in the British Isles. *Wildfowl* **27**, 143–147.

Owen, M. (1977*a*) *Wildfowl of Europe*. London, Macmillan.

Owen, M. (1977*b*) The role of refuges on agricultural land in lessening the conflict between farmers and geese in Britain. *Biol. Conserv.* **11**, 209–222.

Owen, M. (1980*a*) *Wild Geese of the World.* London, Batsford.

Owen, M. (1980*b*) The role of refuges in wildfowl management. pp. 144–156 in Wright, E.N., Inglis, I.R. and Feare, C.J. (eds), *Bird Problems in Agriculture* Croydon, BCPC Publications.

Owen, M. (1982) Population dynamics of Svalbard Barnacle Geese 1970–80: the rate, pattern and causes of mortality as determined by individual marking. *Aquila* **89**, 229–247.

Owen, M. (1990) The damage-conservation interface illustrated by geese. *Ibis.*

Owen, M. and Black, J.M. (1989*a*) Factors affecting the survival of Barnacle Geese on migration from the breeding grounds. *J. Anim. Ecol.* **58**, 603–618.

Owen, M. and Black, J.M. (1989*b*) The Barnacle Goose. pp. 349–362 in Newton, I. (ed), *Lifetime Reproduction in Birds.* Oxford, Blackwell Scientific Publications.

Owen, M. and Black, J.M. (1990) The significance of migration mortality in non-passerine birds, in press, in Perrins, C.M. (ed.), *Proc. Symp. on Population Dynamics and their relevance to Conservation.* The Camargue, France, December 1988.

Owen, M. and Cadbury, C.J. (1975) The ecology and mortality of swans at the Ouse Washes, England. *Wildfowl* **26**, 31–42.

Owen, M. and Cook, W.A. (1977) Variations in body weight, wing length and condition of Mallard *Anas platyrhynchos platyrhynchos* and their relationship to environmental changes. *J. Zool. Lond.* **183**, 377–395.

Owen, M and Dix, M. (1986) Sex ratios in some common British wintering ducks. *Wildfowl* **37**, 104–112.

Owen, M and Gullestad, N. (1984) Migration routes of Svalbard Barnacle Geese *Branta leucopsis* with a preliminary report on the importance of the Bjornoya staging area. *Norsk Polarinstisutt Skrifter* **181**, 67–77.

Owen, M. and King, J.G. (1981) The duration of the flightless period in freeliving Mallard. *Bird Study* **27**, 267–269.

Owen, M. and Mitchell, C.R. (1988) Movements and migration of Wigeon *Anas penelope* wintering in Britain and Ireland. *Bird Study* **35**, 47–49.

Owen, M. and Montgomery, S (1978) Body measurements of Mallard caught in Britain. *Wildfowl* **29**, 123–134.

Owen, M. and Norderhaug, M. (1977) Population dynamics of Barnacle Geese *Branta leucopsis* breeding in Svalbard, 1948–1976. *Ornis Scand.* **8**, 161–174.

Owen, M. and Ogilvie, M.A. (1979) Wing molt and weights of Barnacle Geese in Spitsbergen. *Condor* **81**, 42–52.

Owen, M. and Thomas, G.J. (1976) The feeding ecology and conservation of Wigeon wintering in the Ouse Washes, England. *J. Appl. Ecol.* **16**, 795–809.

Owen, M. and West, J. (1988) Variation in egg compositon in semi-captive Barnacle Geese. *Ornis Scand.* **19**, 58–62.

Owen M., Atkinson-Willes, G.L. and Salmon, D.G. (1986) *Wildfowl in Great Britain*. Second Edition. Cambridge, University Press.

Owen, M., Black, J.M. and Liber, H. (1988) Pair bond duration and timing of its formation in Barnacle Geese (*Branta leucopsis*). pp. 257–269 in Weller, M.W. (ed.), *Waterfowl in Winter* Minneapolis, University of Minnesota Press.

Owen, M., Drent, R.H., Ogilvie, M.A. and van Spanje, T.M. (1978) Numbers, distribution and catching of Barnacle Geese (*Branta leucopsis*) on the Nordenskioldkysten, Svalbard, in 1977. *Norsk Polarinstitutt Aarbok* 1977, 247–258.

Patterson, I.J. (1982) *The Shelduck*. Cambridge, University Press.

Patternson, I.J. (1990) Conflict between geese and agriculture; does goose grazing cause damage to crops? *Ardea* **78**, in press.

Patterson, J.H. (1979) Experiences in Canada. *Trans. 44th N. Amer. Wildl. and Nat. Resources Conf. 1979*, 114–126.

Paulus, S.L. (1984) Activity budgets of non-breeding Gadwalls in Louisiana. *J. Wildl. Manage.* **48**, 371–380.

Pedroli, J.-C. (1982) Activity and time budgets of Tufted ducks on Swiss lakes during winter. *Wildfowl* **33**, 105–112.

Percival, S. (1988) *Grazing Ecology of Barnacle Geese* Branta leucopsis *on Islay*. Unpubl. PhD Thesis, University of Glasgow.

Percival, S. in press. Population structure of Greenland Barnacle Geese, *Branta leucopsis* on the wintering grounds on Islay. *J. Appl. Ecol.*

Perdeck, A.C. and Clason, C. (1983) Sexual differences in migration and winter quarters of ducks ringed in the Netherlands. *Wildfowl* **34**, 137–143.

Perrins, C.M. and Birkhead, T.R. (1983) *Avian Ecology*. Glasgow, Blackie.

Perrins, C.M. and Ogilvie, M.A. (1981) A study of the Abbotsbury Mute Swans. *Wildfowl* **32**, 35–47.

Phillipona, J. (1966) Geese in cold winter weather. *Wildfowl Trust Ann. Rep.* **17**, 95–97.

Pienkowski, M.W. and Evans, P.R. (1982) Clutch parasitism and nesting interference between Shelducks at Aberlady Bay. *Wildfowl* **33**, 159–163.

Pienkowski, M.W., Ferns, P.N., Davidson, N.C. and Worrall, D.H. (1984) Balancing the budget: measuring the energy intake and requirements of shorebirds in the field. pp. 29–56 in Evans, P.R., Goss-Custard, J.D. and Hale, W.G. (eds), *Coastal Waders and Wildfowl in Winter*, Cambridge, University Press.

Pitman C.R.S. (1965) The nesting and some other habits of *Alopochen, Plectopterus* and *Sarkidiornis*. *Wildfowl Trust Ann. Rep.* **16**, 115–121.

Poysa, H. (1983) Morphology-mediated niche organization in a guild of dabbling ducks. *Ornis Scandinavica* **14**, 317–326.

Poysa, H. (1984) Temporal and spatial dynamics of waterfowl populations in a wetland area—a community ecological approach. *Ornis Fennica* **61**, 99–108.

Poysa, H. (1987*a*) Numerical and escape responses of foraging teals *Anas crecca* to predation risk. *Bird Behaviour* **71**, 87–92.

Poysa, H. (1987*b*) Costs and benefits of group foraging in the Teal. *Behaviour* **103**, 123–140.

Prevett, J.P. and MacInnes, C.D. (1980) Family and other social groups in Snow Geese. *Wildl. Monogr.* **71**, 1–46.

Prins, H.H.Th. and Ydenberg, R.C. (1985) Vegetation growth and seasonal habitat shift of the barnacle goose (*Branta leucopsis*). *Oecologia* **66**, 122–125.

Prins, H.H.Th., Ydenberg, R.C. and Drent, R.H. (1980) The interaction of Brent Geese *Branta bernicla* and Sea Plantain *Plantago maritima* during spring staging: field observations and experiments. *Acta Bot. Neerl.* **29**, 585–596.

Prop, J. and Loonen, M. (1989) Goose flocks and food exploitation: the importance of being first. pp. 1878–1887 in Ouellet, H. (ed.), *Acta XIX Cong. Int. Om. Vol. II.* Ottawa, University of Ottawa Press.

Prop, J., van Eerden, M. and Drent, R.H. (1984) Reproductive success of the Barnacle Goose in relation to food exploitation on the breeding grounds, western Spistbergen. *Norsk. Polarinstitutt Skrifter* **181**, 87–117.

Pulliam, H.R. and Caraco, T. (1984) Living in groups: is there an optimal group size? pp. 122–147 in Krebs, J.R. and Davies, N.B. *Behavioural Ecology: An Evolutionary Approach.* 2nd Edition. Oxford, Blackwell Scientific Publications.

Pulliam, H.R. and Millikan, G.C. (1982) Social organization in the nonreproductive season. pp. 169–197 in Farner, D.S., King, J.R. and Parkes, K.C. (eds), *Avian Biology* Vol V. New York, Academic Press.

Ratcliffe, L., Rockwell, R.F. and Cooke, F. (1988) Recruitment and maternal age in Lesser Snow Geese. *J. Anim. Ecol.* **57**, 553–563.

Raveling, D.G. (1969) Social classes of Canada Geese in winter. *J. Wildl. Manage.* **33**, 304–318.

Raveling, D.G. (1979) Traditional use of migration and winter roost sites by Canada Geese. *J. Wildl. Manage.* **43**, 229–235.

Raveling, D.G. (1989) Nest predation rates in relation to colony size of Black Brant. *J. Wildl. Manage.* **53**, 87–90.

Rayner, J.M.V. (1985) Flight, speeds of. pp. 224–226 in Campbell, B. and Lack, E. (eds), *A Dictionary of Birds.* Calton, T. and Poyser, A.D.

Rees, E.C. (1982) The effect of photoperiod on the timing of spring migration in the Bewick's Swan. *Wildfowl* **33**, 119–132.

Rees, E.C. (1987) Conflict of choice within pairs of Bewick's Swans regarding their migratory movement to and from the wintering grounds. *Anim. Behav.* **35**, 1685–1693.

Rees, E.C. and Hillgarth, N. (1984) The breeding biology of captive Black-headed Ducks and the behaviour of their young. *Condor* **86**, 242–250.

Reeves, H.M., Dill, H.H. and Hawkins, A.S. (1968) A case study in Canada Goose management: the Mississippi Valley Population. pp. 150–165 in Hine, H.L. and Schoenfeld, C. (eds), *Canada Goose Management* Madison, Dembar Educational Inc.

Reynolds, C.M. (1972) Mute Swan weights in relation to breeding. *Wildfowl* **23**, 111–118.

Ridgill, S.R. and Fox, A.D. (1989) Cold weather movements of waterfowl in western Europe. Unpubl. Rep. Wildfowl and Wetlands Trust.

Rockwell, R.F., Findlay, C.S. and Cooke, F. (1983) Life history studies of the Lesser Snow Goose (*Anser caerulescens caerulescens*). I. The influence of age and time on fecundity. *Oecologia (Berlin)* **56**, 318–322.

Rogers, J.P. (1964) Effect of drought on reproduction of Lesser Scaup. *J. Wildl. Manage.* **28**, 213–222.

Rogers, J.P. (1979) *Branta bernicla hrota* in the USA—a management review. pp. 198–212 in Smart, M. (ed.), *Proc. First Tech. Meeting on Western Palearctic Migratory Bird Management Paris 1977.* Slimbridge, IWRB.

Rohwer, F.C. (1985) The adaptive significance of clutch size in prairie ducks. *Auk* **102**, 354–361.

Rohwer, F.C. and Anderson, M.G. (1988) Female biased philopatry, monogamy, and the timing of pair formation in waterfowl. pp. 187–214 in Johnston, R.F. (ed.), *Current Ornithology*. New York, Plenum Press.

Ruger, A. (1985) (ed.), *Extent and control of goose damage to agricultural crops. IWRB Special Publ. No. 5.* Slimbridge, IWRB.

Ryder, J.P. (1967) The breeding biology of the Ross' Goose in the Perry River Region, Northwest Territories. *Can. Wildl. Serv. Report Series* **3**, 56 pp.

Ryder, J.P. (1970) A possible factor in the evolution of clutch size in Ross' Goose. *Wilson Bulletin* **82**, 5–13.

St. Joseph, A.K.M. (1979) The seasonal distribution and movements of *Branta bernicla* in western Europe. pp. 45–59 in Smart, M. (ed.), *Proc. 1st Tech. Mtng. on Migratory Bird Management, Paris 1977*. Slimbridge, IWRB.

Salomonsen, F. (1968) The moult migration. *Wildfowl* **19**, 5–24.

Sanderson, G.C. (1978) Conservation of Waterfowl. pp. 43–58 in Bellrose, F.C. 1978. *Ducks, Geese and Swans of North America. Second Edition.* Harrisburg, Stackpole Books.

Savard, J-P.L. (1980) Variability of waterfowl counts in the Cariboo Parkland, British Columbia. pp. 107–122 in Miller, F.L. and Gunn, A. (eds). *Proc. Symp. on Census and Inventory Meth.* Idaho, Forest, Wildl. & Range Expt. Sta.

Savard, J-P.L. (1985) Evidence of long term pair bonds in Barrow's Goldeneye *Bucephala islandica*. *Auk* **102**, 389–391.

Savard, J-P.L. (1988) Winter, spring and summer territoriality in Barrow's Goldeneye: characteristics and benefits. *Ornis Scand.* **19**, 119–128.

Savard, J-P.L. and Smith, J.N.M. (1987) Interspecific aggression by Barrow's Goldeneye: a descriptive and functional analysis. *Behaviour* **102**, 168–184.

Sayler, R.D. and Afton, A.D. (1981) Ecological aspects of Common Goldeneyes *Bucephala clangula* wintering on the upper Mississippi River, U.S.A. *Ornis Scand.* **12**, 99–108.

Schindler, M. and Lamprecht, J. (1987) Increase of parental effort with brood size in a nidifugous bird. *Auk* **104**, 688–693.

Sedinger, J.S. and Raveling, D.G. (1986) Timing of nesting by Canada Geese in relation to the phenology and availability of their food plants. *J. Anim. Ecol.* **55**, 1083–1102.

Scott, D.A. (1980) A preliminary inventory of wetlands of international importance for waterfowl in west Europe and northwest Africa. *IWRB Special Publ No. 2.* Slimbridge, IWRB.

Scott, D.A. (1982) Problems in the management of waterfowl populations. pp. 89–106 in Scott, D.A. and Smart, M. (eds), *Proc. 2nd Tech. Meeting on Western Palearctic Migratory Bird Management. Paris 1979.* Slimbridge, IWRB.

Scott, D.A. (1989) *A Directory of Asian Wetlands.* Gland, Switzerland, IUCN.

Scott, D.A. (1989) Proposal for a coding system for wetland types. *Unpubl. Report*, Slimbridge, Ramsar Convention Secretariat.

Scott, D.A. and Carbonell, M. (1986) A Directory of Neotropical Wetlands. Cambridge, IUCN/IWRB.

Scott, D.K. (1980) Functional aspects of prolonged parental care in Bewick's Swans. *Anim. Behav.* **28**, 938–952.

Scott, D.K. (1984) Winter territoriality of Mute Swans *Cygnus olor*. *Ibis* **126**, 168–176.

Scott, D.K. (1988) Breeding success of Bewick's Swans. pp. 220–236 in Clutton-Brock, T.H. (ed.), *Reproductive Success.* Chicago, University of Chicago Press.

Sears, J. (1988) A report on lead poisoning in Mute Swans in the Thames area during 1988. *Unpubl. Rep.* University of Oxford.

Shillinger, J.E. and Cottam, C.C. (1937) The importance of lead poisoning in waterfowl. *Trans. North Amer. Wildl. Conf.* **2**, 398–403.

Siegfried, W.R. (1970) Wildfowl distribution, conservation and research in southern Africa. *Wildfowl* **21**, 89–98.

Siegfried, W.R. (1976) Social organisation in Ruddy and Maccoa Ducks. *Auk* **93**, 560–570.

Siegfried, W.R. (1977) Notes on the behaviour of Ruddy Ducks during the brood period. *Wildfowl* **28**, 126–128.

Siegfried, W.R. (1979) Social behaviour of the African Comb Duck. *Living Bird* **17**, 85–104.

Siegfried, W.R. (1985) Socialy-induced suppression of breeding plumage in the Maccoa Duck. *Wildfowl* **36**, 135–137.

Sjoberg, K. (1985) Foraging activity patterns in the Goosander (*Mergus merganser*) and the Red-breasted Merganser (*M. serrator*) in relation to patterns of activity in their major prey species.

Sjoberg, K. (1988) Food selection, food-seeking patterns and hunting success of captive Goosanders *Mergus merganser* and Red-breasted Mergansers *M. serrator* in relation to the behaviour of their prey.

Smart, G. (1965) Development and maturation of primary feathers of Redhead ducklings. *J. Wildl. Manage.* **29**, 533–536.

Smith, G.I. (1970) Response of Pintail breeding populations to drought. *J. Wildl. Manage.* **34**, 943–946.

Smith, G.R. (1976) Botulism in waterfowl. *Wildfowl* **27**, 129–138.

Solman, V.E.F. (1945) The ecological relations of Pike, *Esox lucius* and waterfowl. *Ecology* **26**, 157–170.

Sowls, L.K. (1955) *Prairie Ducks*. Washington D.C. Wildlife Management Institute.

Spray, C.J. (1981) An isolated population of *Cygnus olor* in Scotland. pp. 191–208 in Matthews, G.V.T. and Smart, M. (eds), *Proc. 2nd Swan Symp., Sapporo, Japan, 1980*. Slimbridge, IWRB.

Stewart, A.G. (1978) Swans flying at 8,000 metres. *Brit. Birds* **71**, 459–460.

Stirling, T. and Dzubin, A. (1967) Canada Goose moult migration to the North-west Territories. *Trans. N. Amer. Wildlife and Nat. Resources Conf.* **31**, 355–373.

Stoudt, I.J. and Cornwell, G.W. (1976) Nonhunting mortality of North American waterfowl. *J. Wildl. Manage.* **40**, 681–693.

Strange, T.H., Cunningham, E.R. and Goertz, J.W. (1971) Use of nest boxes by Wood Ducks in Mississippi. *J. Wildl. Manage.* **35**, 786–793.

Street, M. (1978) The role of insects in the diet of Mallard ducklings—an experimental approach. *Wildfowl* **29**, 93–100.

Summers, R.W. (1982) The absence of flightless moult in the Ruddy-headed Goose in Argentina and Chile. *Wildfowl* **33**, 5–6.

Summers. R.W. (1983) Moult-skipping by Upland Geese *Chloephaga picta* in the Falkland Islands. *Ibis* **125**, 262–266.

Summers, R.W. and Dunnett, G. (1984) Sheldgeese and Man in the Falkland Islands. *Biol. Cons.* **30**, 319–340.

Summers, R.W. and Grieve, A. (1982) Diet, feeding behaviour and food intake of the Upland Goose (*Chloephaga picta*) and Ruddy-headed Goose (*C. rubidiceps*) in the Falkland Islands. *J. Appl. Ecol.* **19**, 783–804.

Summers, R.W. and Underhill, L.G. (1987) Factors related to breeding production of Brent Geese *Branta bernicla* and waders (Charadrii) on the Taimyr Peninsula. *Bird Study* **33**, 105–108.

Swanson, G.A., Shaffer, T.L., Wolf, J.F. and Lee, F.B. (1986) Renesting characteristics of captive Mallards on experimental ponds. *J. Wildl. Manage.* **50**, 32–38.

Tamisier, A. (1972) Rhythmes nychtemeraux des sarcelles d'hiver pendant leur hivernage en Camargue. *Alauda* **48**, 107–135.

Tamisier, A. (1974) Etho-ecologicl studies of Teal entering in the Camargue (Rhone Delta, France). *Wildfowl* **25**,•107–117.

Tamisier, A. (1985) Some considerations on the social requirements of ducks in winter. *Wildfowl* **36**, 104–108.

Taylor, J. (1953) A possible moult migration of Pink-footed Geese. *Ibis* **95**, 638–642.

Teunissen, W., Spaans, B. and Drent, R.H. (1985) Breeding success in Brent Geese in relation to individual feeding opportunities during spring in the Wadden Sea. *Ardea* **73**, 109–119.

Thomas, C.B. (1977) The mortality of Yorkshire Canada Geese. *Wildfowl* **28**, 35–47.

Thomas, G.J. (1981) Field feeding by dabbling ducks around the Ouse Washes, England. *Wildfowl* **32**, 69–78.

Thompson, S.C. and Raveling, D.G. (1987) Incubation behaviour of Emperor Geese compared with other geese: interactions of predation, body size, and energetics. *Auk* **104**, 707–716.

Thompson, S.C. and Raveling, D.G. (1988) Nest insulation and incubation constancy of arctic geese. *Wildfowl* **39**, 124–132.

Torres Esquivias, J.A. and Ayala Moreno, J.M. (1986) Variation idu dessein cephalique des males de l'erismature à tête blanche (*Oxyura leucocephala*). *Alauda* **54**, 197–206.

Trivers, R.L. (1972) Parental investment and sexual selection. pp. 136–179 in Campbell, B. (ed.), *Sexual selection and the descent of Man.* Chicago, Aldine Publ. Co.

Tuite, C.H., Owen, M. and Paynter, D. (1983) Interaction between wildfowl and recreation at Llangorse Lake and Talybont Reservoir, South Wales. *Wildfowl* **34**, 48–63.

Van der Waal, R.J. and Zomerdijk, P.J. (1979) The moulting of Tufted Duck and Pochard in the Ijsselmeer in relation to moult concentrations in Europe. *Wildfowl* **30**, 99–108.

Van Eerden, M.R. (1984) Waterfowl movements in relation to food stocks. pp. 85–100 in Evans, P.R., Goss-Custard, J.D. and Hale, W.G. (eds), *Coastal waders and wildfowl in winter* Cambridge, University Press.

Van Eerden, M.R. (1990) Moulting Greylag Geese in the Netherlands and their effect on the vegetation. *Ardea* **78.**

Van Welie, H.W.M. (1985) Compensation for goose damage in the Netherlands. pp. 29–33 in Ruger, A. (ed.), *Extent and control of goose damage to agricultural crops. IWRB Special Publ. No. 5.* Slimbridge, IWRB.

Vaught, R.W. and Kirsch, L.M. (1966) Canada Geese of the eastern prairie population, with special reference to the Swan Lake flock. *Missouri Dept. of Conservation Tech. Bull.* **3**, 91 pp.

Vaught, R.W., McDougle, H.C. and Burgess, H.H. (1967) Fowl Cholera in waterfowl at Squaw Creek National Wildlife Refuge, Missouri. *J. Wildl. Manage.* **31**, 248–253.

Vehrencamp, S.L. and Bradbury, J.W. (1984) Mating systems and ecology. pp. 251–303 in Krebs, J.R. and Davies, N.B. *Behavioural Ecology: An Evolutionary Approach.* 2nd Edition. Oxford, Blackwells Scientific Publications.

Ward, P. and Zahavi, A. (1973) The importance of certain assemblages of birds as 'information centres' for food finding. *Ibis* **115**, 517–534.

Weller, M.W. (1959) Parasitic egg laying in the Redhead (*Aythya americana*) and other North American Anatidae. *Ecol. Monogr.* **29**, 333–365.

Weller, M.W. (1965) Chronology of pair formation in some Nearctic *Aythya* (Anatidae). *Auk* **82**, 227–235.

Weller, M.W. (1968) The breeding biology of the parasite Black-headed Duck. *Living Bird* **7**, 167–207

Weller, M.W. (1976) Ecology and behaviour of Steamer Ducks. *Wildfowl* **27**, 45–53.

Weller, M.W. (1980) *The Island Waterfowl.* Ames, Iowa, Iowa State University Press.

Welty, J.C. and Baptista, L. (1988) *The Life of Birds*, 4th Edition. New York, Saunders.

Williams. G.C. (1966) Natural selection, and the costs of reproduction, a refinement of Lack's hypothesis. *Amer. Nat.* **100**, 687–690.

Williams, M. (1973) Mortality of the Black Swan in New Zealand—a progress report. *Wildfowl* **24**, 94–102.

Williams, M.J. (1975) Creching behaviour of the Shelduck. *Ornis Scand.* **5**, 131–143.

Wilson, H.J., Norriss, D.W., Walsh, A., Fox, A.D. and Stroud, D.A. (1990) Winter site fidelity in Greenland White-fronted Geese: implications for conservation and management. *Ardea* **78.**

Wood, C.C (1986) Dispersion of Common Merganser (*Mergus merganser*) breeding pairs in relation to the availability of juvenile Pacific Salmon in Vancouver Island streams. *Can. J. Zool.* **64**, 756–765.

Wood, C.C. (1987) Predation of juvenile Pacific Salmon by the Common Merganser *Mergus merganser* on eastern Vancouver Island. II. Predation of stream-resident juvenile salmon by Merganser broods. *Can. J. Zool.* **64**, 950–959.

Wood, C.C. and Hand, C.M. (1985) Food-searching behaviour of the Common Merganser (*Mergus merganser*). I. Functional responses to prey and predator density. *Can J. Zool.* **63**, 1260–1270.

Woolington, D.W., Stringer, P.F. and Yappaguire, D.R. (1979) Migration and distribution of Aleutian Canada Geese. pp. 299–309 in Jarvis, R.L. and Bartonek, J.C. (eds), *Management and Biology of Pacific Flyway Geese* Corvallis, OSU book Stores Inc.

Wypkema, R.C.P. and Ankney, C.D. (1979) Nutrient reserve dynamics of Lesser Snow Geese staging at James Bay, Ontario. *Can. J. Zool.* **57**, 213–219.

Ydenberg, R.C., Prins, H.H.Th. and van Dijk, J. (1983) The post-roost gatherings of wintering Barnacle Geese: information centres? *Ardea* **71**, 125–131.

Young, C.M. (1970) Territoriality in the Common Shelduck *Tadorna tadorna. Ibis* **112**, 330–335.

Index